RED CLOVER SCIENCE

Current Plant Science and Biotechnology in Agriculture

VOLUME 28

Aims and Scope
The book series is intended for readers ranging from advanced students to senior research scientists and corporate directors interested in acquiring in-depth, state-of-the-art knowledge about research findings and techniques related to all aspects of agricultural biotechnology. Although the previous volumes in the series dealt with plant science and biotechnology, the aim is now to also include volumes dealing with animals science, food science and microbiology. While the subject matter will relate more particularly to agricultural applications, timely topics in basic science and biotechnology will also be explored. Some volumes will report progress in rapidly advancing disciplines through proceedings of symposia and workshops while others will detail fundamental information of an enduring nature that will be referenced repeatedly.

The titles published in this series are listed at the end of this volume.

Red Clover Science

by

N.L. TAYLOR
University of Kentucky, Agronomy,
Lexington, Kentucky, USA

and

K.H. QUESENBERRY
University of Florida, Agronomy Department,
Gainesville, Florida, USA

KLUWER ACADEMIC PUBLISHERS
DORDRECHT / BOSTON / LONDON

Library of Congress Cataloging-in-Publication Data

Taylor, N. L. (Norman L.)
Red clover science / by N.L. Taylor and K.H. Quesenberry.
p. cm. -- (Current plant sciences and biology in agriculture ; v. 28)
ISBN 0-7923-3887-1 (hardcover : alk. paper)
1. Red clover. I. Quesenberry, K. H. II. Title. III. Series: Current plant sciences and biology in agriculture ; 28.
SB205.C64T39 1996
634.3'2--dc20 95-51138

ISBN 0-7923-3887-1

Published by Kluwer Academic Publishers,
P.O. Box 17, 3300 AA Dordrecht, The Netherlands

Kluwer Academic Publishers incorporates
the publishing programmes of
D. Reidel, Martinus Nijhoff, Dr W. Junk and MTP Press.

Sold and distributed in the U.S.A. and Canada
by Kluwer Academic Publishers,
101 Philip Drive, Norwell, MA 02061, U.S.A.

In all other countries, sold and distributed
by Kluwer Academic Publishers Group,
P.O. Box 322, 3300 AH Dordrecht, The Netherlands

Printed on acid-free paper

Printed in the Netherlands

Contents

Reviewers

1. Historical Perspectives
   a. Jan Sjodin. Svalof AB, S-268 Svalov, Sweden
2. Biosystematics and Interspecific Hybridization
   a. Jude Grosser, Citrus Research and Education Center, Lake Alfred, Florida, USA
3. Reproductive Biology, Genetics, and Evolution
   a. Wayne Parrott, University of Georgia, Athens, GA, USA
4. Morphology and Physiology
   a. W.A. Kendall, USDA Pasture Laboratory, University Park, PA, USA
5. Management, Utilization, Quality, and Antiquality
   a. M. Collins, Univ. of Kentucky, Lexington, KY, USA
6. Fungal Diseases
   a. Craig Grau, University of Wisconsin, Madison, WI, USA
7. Virus Diseases
   a. S.A. Ghabrial, University of Kentucky, Lexington, KY, USA
8. Nematode Diseases
   a. R.A. Skipp, Agr. Research, Grasslands, Palmerston North, New Zealand
9. Insect Pests
   a. K.V. Yeargan, Univ. of Kentucky, Lexington, KY, USA
10. Persistence
    a. Bengt Andersson, Svalof AB, S-268, Svalov, Sweden
    b. Peter Marum, Loken Research Station, Norway
    c. C.T. Dougherty, University of Kentucky, Lexington, KY, USA
11. Rhizobium Relationships
    a. Eric Triplett, Univ. of Wisconsin, Madison, WI, USA
12. Breeding Methodology
    a. Clair Mousset-Declas, Inst. National de la Recherche Agronomique, Centre Recherche de Dijon, Station d'Ameliortion des Plantes, Dijon, France
13. Tetraploid Red Clover
    a. Jan Sjodin, Svalof AB, S-268 Svalov, Sweden

14. Tissue Culture
    a. Greg Phillips, New Mexico State University, Las Cruces, NM, USA
15. Seed Production
    a. William Rumball, Ag Research Grasslands, Palmerston North, New Zealand
    b. T.M. Choo, Seed Division, Agric. Canada, Ottawa
    c. Real Michaud, Seed Division, Agric. Canada, Quebec City
16. Germplasm Acquisition Maintenance and Evaluation
    a. Warren Williams, Margot Forde Germplasm Centre, Palmerston North, New Zealand
17. Future Trends
    a. Ed Perfect, University of Kentucky, Lexington, KY, USA

Preface

Red clover (*Trifolium pratense* L.) is an important forage of most of the temperate regions of the world, and at one time was the most important legume for hay in the United States. Oddly enough, red clover has never been the subject of a monograph, although its culture has been discussed in many monograph chapters in the United States and Europe, perhaps beginning with those of Westgate and Hillman in a USDA Farmers Bulletin in 1911 and early Swedish publications. The monograph "Clover Science and Technology" published in 1985 provides coverage of red clover along with other *Trifolium* species. It was the objective of the present monograph, "Red Clover Science" to examine the literature published on red clover primarily after 1984. Perhaps the timing of this publication is appropriate with the present interest in sustainable agriculture of which red clover was so prominent in the past. Red clover and other forage legumes are valuable for reducing our reliance on fossil fuels, and for reducing pollution of our soil, water, and atmosphere primarily from use of chemical nitrogen and other fertilizers. In each chapter, an attempt was made to summarize the earlier information and to integrate the recent findings into this background. References earlier than 1984 are included only when necessary for historical purposes. No attempt was made to review the many aspects of clover culture that has been described thoroughly in earlier publications. The literature covered is necessarily worldwide because of present greater emphasis in northern Eurasia than in United States. For this reason, we requested authorities in red clover producing countries to review individual chapters. However, because the authors are U.S. based, it was difficult to avoid some provincialism for which the authors apologize. The readership is expected to be primarily scientists and scientifically trained technicians who will appreciate an up-to-date summary of knowledge on this important species. The authors acknowledge support from the Universities of Kentucky and Florida. The senior author especially wishes to acknowledge the University of Florida for providing office, library, and secretarial help during a sabbatical in 1992–93. The junior author gratefully acknowledges the USDA-ARS National Forage Seed Production Research Center, Corvallis, Oregon, for providing office, library, and secretarial assistance during a sabbatical in 1994. The senior author acknowledges the help of secretaries Tracy Brown and Susan Leopold in typing the many drafts of the chapters, and technician Richard Mundell for proofing and providing helpful suggestions. The junior author acknowledges the help of secretaries Kim Lottinville for typing and revising chapters, and Traci Martin for assistance with reference checking and editing. Finally, the authors acknowledge the many professionals listed under reviewers who reviewed the individual chapters and provided many helpful suggestions.

Norman L. Taylor
Kenneth H. Quesenberry

⟶

Trifolium species related to *T. pratense*.
Left - top to bottom *T. pratense* showing variation in flowers; *T. medium, T. sarosiense*.
Right - top to bottom *T. pallitum* (R), diploid *T. diffusium* (L), tetraploid *T. diffusum*, *T. alpestre*.

1. Historical perspectives

Red clover (*Trifolium pratense* L.) is one of about 250 species of the genus *Trifolium*. Due to high seedling vigor, easy establishment, rapid growth, and soil improving characteristics it became a very important crop first in temperate regions of Europe, later in the United States and finally world wide (Fig. 1.1).

Red clover is thought to have originated in southeastern Eurasia, near the Mediterranean Sea in one of the three major centers of diversity of the genus (See Chapter 2, Biosystematics and Interspecific Hybridization). According to Merkenschlager (1934), wild red clover was distributed over much of western Europe. A large number of regional races were developed that were adapted to a wide variety of areas, bridging all zones from humid to arid climates, and from coastal to continental regions. In southern Europe, red clover became a plant of hill climates, but still was definitely a plant of humid climates.

After a long history as a wild plant, red clover probably first assumed the rank of a cultivated plant in Spain (Merkenschlager, 1934). From there it was introduced and reintroduced into Northern Europe.

Historical events concerning red clover may be listed as follows:

- Cultivated in Europe in the third and fourth centuries.
- First mention of cultivation was by Abertus Magnus in 13th century.
- Recorded as being in Italy by 1550.
- In Flanders by 1566.
- In France by 1585.
- In England by 1645.
- In United States by 1663.
- In Russia by 1776.

Red clover has now spread to most temperate regions of the world; from the Mediterranean Sea to North Scandinavia, and the former USSR in Eurasia; in most of North America from southern Florida and Texas to Canada and Alaska; in South America north from Chile and Argentina to higher elevations in Peru and Mexico; in New Zealand and Australia and in northern China to Japan. It is not adapted to dry or hot humid climates (See Chapter 4, Morphology and Physiology). After its first introduction into most countries,

Fig. 1. 1. Typical red clover plants growing in a mixture with grass.

its usage was initially resisted by farmers, but gradually they realized its value and used it to replace the fallow in agricultural systems.

According to Weir (1926), Flemish farmers abolished the fallow, and from time immemorial had been growing clover in rotations with crops like hemp, turnips, and small grains. Flemish farmers had two proverbial sayings, which still have relevance today: "No forage, no cattle; without cattle, no manure, and without manure, no crop"; and "Without clover no man in Flanders would presume to call himself a farmer." The Flemish farmers were so successful that as early as 1600, Flanders had become the school of husbandry in Europe.

1. Early literature on red clover

As cultivated red clover was introduced into new areas, it crossed with the wild forms and a large number of ecological races were developed. In Germany, an early pioneer in red clover cultivation was J.C. Schubart who in 1783 won the prize of the Berlin Academy of Science for the best essay on "The various properties and the successful growing of forage plants". Eventually, red clover was cultivated in Switzerland where it was known as "Mattenklee" (Merkenschlager, 1934). In Wales, an important advance in the early 1800's was made when studies revealed that local strains of red clover consisted of distinct types. Until that time farmers who used a local strain successfully were convinced that its superior merit was simply due to the fact that the seed was home grown, and did not realize that the **strain** was unique (Evans, 1957).

In the U.S., according to Fergus and Hollowell (1960), red clover seed was advertised for sale on April 10, 1729 in a colonial newspaper. Also in "Narratives of Early Maryland", 1633–1684, settlers were advised to take "good store" of clover grass seed to make good meadow. Doubtless, as colonists moved into new areas, they carried seeds, perhaps in hayloft chaff, or in straw or hay. Scattered references to red clover exist in the early farm literature and scientific literature began to appear in the United States according to the following chronology: Westgate and Hillman (1911) detailed the culture of red clover. No cultivars were known but two types of red clover were listed, ordinary and mammoth. The mammoth type was also known as sapling clover, pea vine clover, bull clover and perennial clover. The ordinary clover was intermediate in earliness to bloom between the English early and the late single cut (mammoth) and thus was called medium or double cut red clover. This term was for a time confused with the binomial for zigzag clover, *Trifolium medium* L. Russian red clover was also described by Westgate and Hillman as not well adapted but, being glabrous was much less dusty than American types. Even at this early date, they listed a "special form", apparently a reference to breeding to develop new cultivars. One to two bushels of seed was thought to be a good seed crop. A method of determining if the crop was worth saving for seed was as follows: If a "goodly" number of heads each have 25 seeds, the crop may be expected to yield one to two bushels of seed. This method of estimation was perpetuated in the literature for many years, and often was used by farmers as rule of thumb to determine whether the crop was worth saving for seed, or should be harvested for hay. At this time, Piper (1916) published the first version of "Forage Plants and Their Culture" in the Rural Text Series which included a complete treatment on methods of growing red clover. This made the information readily available to farmers of the United States. Pieters (1923) revised Westgate and Hillman's bulletin and added some new information. The first cultivar listed was Tennessee Anthracnose Resistant but no seed was available in commerce. A glabrous type was imported from Chile which, like the Russian type, was recognized for its lack of dustiness.

Westgate and Hillman (1911) recognized "clover sickness", or "clover-sick soils". They listed as probable causes; exhaustion of fertilizer, presence of deleterious substances left in the soil by clover plants, clover anthracnose (in Tennessee), lack of phosphorous and lime (in Illinois), nematodes (in West Virginia), and exhaustion of the humus supply. Their remedies were to apply manure, sow other crops, in particular, alsike clover (*T. hybridum* L.). After an interval of 5 to 8 years it became possible to grow clover again.

Pieters (1924) wrote a farmers bulletin that dealt exclusively with "clover failure" a term he advocated to replace "clover sickness". Clover failure had been very prevalent in the previous 40 years, and farmers were concerned as to the remedy. New theories advanced as to its cause in Europe and in America were: Excreta of insects poisoned the soil, and/or toxins excreted by the plant itself were responsible. Pieters stated that the causes of clover failure were:

Soil unsuited.
Attacks by a fungus, insect or nematode.
Unadapted cultivars.
Poor methods of seeding.
Wrong fall treatment the first year.

Pieters and Hollowell (1931) emphasized that clover failure in the southern part of the clover belt was associated with susceptibility to southern anthracnose (B & E). Selections for resistance to this disease were conducted much earlier by Bain and Essary (1906) who were also the naming authorities for the causal fungus. Studies aimed at elucidating the cause of clover failure in Kentucky were initiated by Fergus (1931). Possible causes of failure included the diseases southern anthracnose, northern anthracnose (caused by *Kabatiella caulivora* (Kirchn.) Karak., crown rot (caused by *Slerotinia trifoliorum* Erik.), root rot (causal agent unknown), black stem disease (causal agent not given), *Rhizoctonia*, leafhopper injury (caused by *Empoasca fabae* Harris), winterkilling, drought and low soil productivity.

Fergus concluded that the solution to Kentucky's clover failure problem was to use varieties that had become adapted to Kentucky conditions by years of natural selection within the state. None of the known causes of failure, except low soil productivity and root rot, appreciably injured the adapted varieties.

By 1937, considerable research on the clovers had been conducted and was summarized by Pieters and Hollowell (1937) in the USDA yearbook. They pointed out that red clover is such a variable plant that the many cultivars that had developed since the introduction of red clover into England in 1645 resulted from the action of natural selection under local conditions rather than conscious selection by man. For example, in The Netherlands, a cultivar, Maas, was characterized by the almost total absence of the "V" shaped white spot on the leaflets, and in the U.S. the rough hairiness distinguished the American clover from its European progenitor. They surmised that this hairiness resulted from attacks of leafhoppers which prevented seed production in smooth plants more than hairy plants.

Beginning in 1928, the USDA began a search for red clover stocks that had been grown continuously on the same farms for long periods. Seeds of approximately 75 strains were secured, many of which had been grown continuously for 40 years. These were sown in observational plots in cooperation with the Kentucky, Ohio, and Iowa Experiment Stations. Seed of these farmer lots have largely disappeared from commerce but the names are still available from old records. They include Stevens, Sanford, Libel, Wegener, Emerson, Bayne, Reinholdt, Ky 215, VanAtta, Rahn, Port Gibson, and Van Fossen. These farmers of the United States, and other countries performed a valuable service in perpetuating clover germplasm until public plant breeding became active. As a result of these studies, three regions of adaptation were recognized in the humid eastern part of the United States, namely southern, central, and northern. The idea was to develop cultivars from the farmer strains that were adapted to the three regions. Research to that end was initiated at various experiment stations.

In other countries, research on cultivar development also was well underway. In Canada, the double cut cultivars, Ottawa and Dollard; and the single cut cultivars, Altaswede and Manhardy were developed. In England, one improved cultivar was S-123, believed useful for 2 and 3-year meadows. In Denmark, the Otofte Early, Otofte Semilate, and Otofte Late were bred to fit various areas and uses. In Sweden, Svalof Purebred, and Merkur were valued for resistance to *Sclerotinia* and stem nematodes (*Ditylenchus*). Cultivars were also developed in Czechoslovakia and Russia.

By this time, it was known that red clover was self sterile and that bees were required for crossing and seed production. The early work of RD Williams (1931) at the Welsh Plant Breeding Station was particularly notable. He devoted much attention to the subject of fertility and sterility in red clover, and devised three methods of breeding; strain building, sib mating, and diallel crossing.

Improved U.S. red clover cultivars developed shortly after World War II years were Midland, originated from a composite of equal proportions of four old strains, one each from Ohio, Indiana, Illinois, and Iowa; and Cumberland, originated from a composite of three strains, one each from Kentucky, Tennessee and Virginia (Hollowell, 1947).

In 1947, the new southern anthracnose resistant cultivar Kenland was released by the Kentucky Agricultural Experiment Station in cooperation with the USDA (Hollowell, 1951). The cultivar was developed by compositing seed of seven farmer strains, sowing, and creating an artificial epiphytotic for the fungus causing southern anthracnose. After elimination of susceptible plants, recurrent selection was practiced for several generations.

In the post-war years, development of red clover was increasingly emphasized. Whyte et al. (1953) listed and described many of the best-known European strains as follows:

A. Double-cut clover
   Silesian - Central European
   Broad red or cowgrass - British
   Mattenklee - Swiss
   Gendringshe Rood Klaver - Dutch
   Otofte early - Denmark
   Essi - Swedish
   Prinshof Selection - South African

B. Medium red clover
   Vale of Clwyd - British
   Dorset Marl - British
   S-151 - Welsh
   Karaby - Swedish
   Merkur - Swedish
   Resistenta - Swedish

C. Single-cut clover
   Montgomery late - British
   S-123 - British
   Gota - Swedish
   Ultuna - Swedish
   Molstad - Norwegian
   Tammisto - Finnish

2. Seed certification and foundation seed projects

In the United States, as well as other countries, one of the goals of agricultural agencies was to supply high quality seed of improved cultivars to farmers. The enactment of Crop Improvement Association programs did much to insure that seed of high quality was distributed. These began as early as 1923 in England and Wales (Kelly and Bowring, 1990). However, by as late as 1945 no reliable means was available for insuring that adequate amounts of newly developed forage cultivars reached their intended target, use by farmers. The principal difficulty was that no mechanism existed whereby seed of improved cultivars could be produced in quantity prior to building up a demand. While experiment stations had produced many new forage cultivars, very few farmers benefitted from them because of a lack of seed.

The National Foundation Seed Program was initiated in the United States as a joint undertaking of the USDA, the seed trade, the state experiment stations, and extension services to increase seed of forage cultivars in western United States. The program was quite successful for red clover as shown by the fact that the Kenland cultivar was increased from 110 metric tons (t) in 1949 to 1359 t in 1954. In 1951, 97.5% of all the Kenland cultivar seed was grown in the Western States. Eighty-five percent of the 1954 production of certified seed was Kenland, of which 80% was grown in Idaho (Wheeler and Hill, 1957; Garrison, 1952). No red clover cultivar has been so dominant in the U.S. seed trade since that time.

The development of the herbage seeds industry in the United Kingdom began about 1940 when seed supplies from other countries were eliminated due to World War II (Bean, 1977). Seed growers' associations were formed

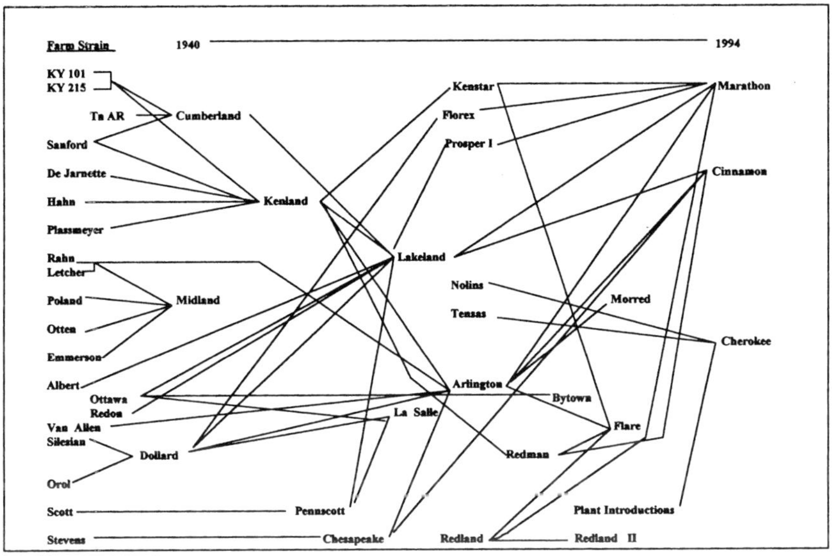

Fig. 1. 2. Source materials used in the breeding of red clover cultivars registered with the American Society of Agronomy and the Crop Science Society of America during 1943 through 1994 (Modified from Rumbaugh, 1991).

which, together with a seed crop inspection and certification scheme, formed the basis of a seed production industry. Authenticity was based on field crop inspection until 1956, when guaranteed seed in sealed bags bearing the warranty of the authority became available. In 1976, the voluntary scheme was changed to a statutory certification system with Britain's entry into the European Economic Community (EEC).

The National Seed Development Organization (NSDO) was formed in 1967 to fulfill the same objectives as the National Foundation Seed Project in the United States; that is, to ensure that seed of cultivars produced by the plant breeder are available in sufficient quantities for farmer use.

The use of a national list of varieties promulgated in 1964 in accordance with the treaty of Rome, required that all EEC countries would provide a list of varieties that were distinct, uniform and stable, and be of value for cultivation. The National List system became effective for Britain in 1973. By 1977, three cultivars of red clover were listed: Sabtoron, Norseman and S-123. All countries of the EEC now have a national list system (Bean, 1977). The number of red clover cultivars eligible for certification is provided by the Organization for Economic Cooperation and Development (OECD). This list, including all cultivars moving in international trade in 1993 is as follows: Diploids, 101, tetraploids, 41, and ploidy unclassified, 11 (Anonymous, 1993).

3. The modern Era

Julen (1959) of Germany and Fergus and Hollowell (1960) of USA summarized the research literature on red clover. Clover sickness was no longer seen as a problem. Instead, the diseases, nematodes and insects attacking red clover were clearly delineated. Morphology was becoming understood, as were the many facets of physiology and ecology. The latest information on breeding, genetics, seed production, and fertilization and management was summarized. Fergus and Hollowell indicated that more research was needed on the effect of high temperature during reproduction, the causes of winter killing, effect of light on biochemistry, relationships of moisture and growth including interactions with pathological and entomological factors, and persistency. These questions and others were addressed in succeeding publications (Taylor and Smith, 1979, 1981; Bowley et al., 1984; and Taylor, 1985).

In the modern era, much more basic research has been undertaken by USDA and Experiment Stations in the United States. However, cultivars have continued to be released by both public and private institutions. The pedigrees of recently released cultivars in the U.S. are shown in Fig. 1.2. The number of cultivars released in U.S. and other countries has greatly increased, and much more emphasis, particularly in Europe has been placed on tetraploid cultivars (See Chapter 13, Tetraploid Red Clover). In the U.S., much research in the public sector is involved with germplasm collection, enhancement, evaluation, and less with cultivar development (See Chapter 16, Germplasm Acquisition, Maintenance and Evalvation).

4. Economic importance

Red clover increased in economic importance in the United States from colonial times, as the eastern states were settled until well into the 20th century. According to Piper (1916) in 1909, red clover totaled almost five million hectares and was grown in every state of the Union and the District of Columbia. Recent federal agricultural census reports and Agricultural Statistics provide no data on red clover grown alone but the combined clover-timothy hectarage can be roughly estimated from data on seed disappearance. According to Fergus and Hollowell (1960), clover grass hectarage appeared to be at its maximum from about 1909 to 1927 during which time it fluctuated between 12 million and 14 million hectares. Between 1927 and 1934, the area declined to about 8 million hectares, and declined again to about 6 million hectares by 1957 (Fergus and Hollowell, 1960). In 1950, according to U.S. Census data, red clover seed production was at a maximum of 67,430 t, but declined to 33,126 t by 1957. Seed production further decreased to 29,366 t in 1965, 19,856 t in 1967, 8,455 t in 1982, and 10,280 t in 1987. The 1987 figure would supply seed for about 4.15 million hectares indicating that red clover usage in the United States is at an all time low. Similar conditions prevail in many other countries in the world including Great Britain, France, Sweden, and New Zealand but apparently the decline has not been as great in Canada (See Chapter 15, Seed Production).

References

Anonymous, 1993. List of cultivars eligible for certification. OECD schemes for the varietal certification of seed moving in international trade. Organization for Economic Cooperation and Development. Paris, France.

Bain, S.M. and S.H. Essary. 1906. Selection for disease resistant clover. *Univ. of Tenn. Agric. Exp. Sta. Bul.* 19:1–10.

Bean, E.W. 1977. Principles of herbage seed production. Welsh Plant Breeding Station publ. In and Ed. pp 1–149. Aberystwyth, Wales.

Bowley, S.R., N.L. Taylor, and C.T. Dougherty. 1984 Physiology and morphology of red clover. *Adv. in Agron.* 37:317–347.

Evans, G. 1957. The clover tradition in Wales. *J. Agric. Univ. Coll. Wales* 28:30–35.

Fergus, E.N. and E.A. Hollowell. 1960. Red clover. *Adv. in Agron.* 12:365–436.

Fergus, E.N. 1931. An analysis of clover failure in Kentucky. *Ky. Agr. Exp. Sta. Bul.* 324:439-476.

Garrison, C.S. 1952. The national foundation seed stocks program in the United States, *Proc. 6th Intern. Grassl. Cong.* 1:918–921.

Hollowell, E.A. 1947. More and better clover. USDA Yearbook. 1943–47:427–432.

Hollowell, E.A. 1951. Registration of varieties and strains of red clover, *II. Jour. Amer. Soc. Agron.* 43:242.

Julen, G. 1959. Rotklee, *Trifolium pratense* L. pp 239–305. In P. Parey (ed.). *Handbuch der Pflanzen-zachtung.* Band IV. Paul Parey in Berlin und Hamburg, Germany.

Kelly, A.F. and J.D.C. Bowring. 1990. The development of seed certification in England and Wales. *Plant Varieties and Seeds.* 3:139–150.

Merkenschlager, F. 1934. Migration and distribution of red clover in Europe. *Herb. Rev.* 2:88–92.

Pieters, A.J. 1923. Red clover culture. *USDA Farmers Bul.* 1339:1–30.

Pieters, A.J. 1924. Clover failure. *USDA Farmers Bul.* 1365:1–24.

Pieters, A.J. and E.A. Hollowell. 1937. *Clover Improvement.* Yearbook of Agric. 1937:1190–1214.

Piper, C.V. 1916. *Forage Plants and Their Culture.* MacMillian, New York.

Rumbaugh, M.D. 1991. Plant Introduction: The foundation of North American forage legume culture development. pp 103–114. In H.L. Shands and L.E. Wiesner (eds.), *Use of Plant Introductions in Cultivar Development.* CSSA Spec. Pub. No. 17. CSSA, Madison, WI.

Taylor, N.L. 1985. Clover Science and Technology. ASA-CSSA-SSA Mono. 25:616 pp. Madison, WI.

Taylor, N.L. and R.R. Smith. 1979. Red clover breeding and genetics. *Adv. in Agron.* 31:125–154.

Taylor, N.L. and R.R. Smith. 1981. Red clover. In T.A. McClure, and E.S. Lipinsky (eds.). *Handbook of Biosolar Resources.* CRC Press. Vol II:11–21. Boca Raton, FL.

Westgate, J.M. and F.H. Hillman. 1911. Red clover. *USDA Farmers Bul.* 455:1–48.

Weir, W.W. 1926. Soil productivity as affected by crop rotation. *USDA Farmers Bul.* 1475:1–22.

Wheeler, W.A. and D.D. Hill. 1957. *Grassland Seeds.* D. Van Nostrand Co., Inc. Princeton, New Jersey.

Whyte, R.O., G. Nilsson-Leissner and H.C. Trumble. 1953. *Legumes in Agriculture.* FAO studies No. 21. Rome, Italy.

Williams, R.D. 1931. Methods and technique of breeding red clover, white clover, and lucerne. *Imp. Bur. Plant. Genetics. Herbage Plants Bul.* 3:46–77.

2. Biosystematics and Interspecific Hybridization

1. Taxonomy

The genus *Trifolium* L. has been divided into eight sections (Zohary and Heller, 1984). Although numerous *Trifolium* species are utilized by grazing animals in their native habitat, only about 11 species are used to any extent in planted pastures. Of these 11, *T. vesiculosum* Savi, arrowleaf clover, is in section *Mistyllus* (C. Presl) Godr., *T. subterraneum* L., sub clover, is in section *Trichocephalum* Koch., and *T. resupinatum* L., persian clover, is in section *Vesicaria* Crantz.. Of the remaining eight species, *T. repens* L., white clover, *T. hybridum* L., alsike clover, *T. nigrescens* Viv., ball clover, and *T. ambiguum* M.B., kura clover are members of section *Lotoidea* Crantz.. The remaining four species, *T. alexandrinum* L., berseem clover, *T. incarnatum* L., crimson clover, *T. hirtum* All., rose clover, and *T. pratense* L., red clover are placed in section *Trifolium* Zoh. *T. pratense* was selected by Zohary and Heller (1984) as the type species (lectotype) of the genus.

Section *Trifolium* is characterized as including both annual and perennial species . Flowering heads may be terminal or axillary and sessile or pedunculate. Flowers are bractless with the calyx being 10–20 nerved and usually hairy. The calyx may have equal or unequal teeth and is usually closed by a callus plug or a hairy ring. The corolla is mostly united and generally persists in fruit. The legume is usually one to two seeded.

Section *Trifolium* consists of 72 species and was divided into 17 subsections by Zohary and Heller (1984). Red clover was placed in subsection *Trifolium* Zoh., along with *T. noricum* Wulf., *T. pallidum* Waldst. & Kit., *T. diffusum* Ehrh., and *T. mazanderanicum* Rech. fil.. Zohary and Heller admit that *T. mazanderanicum* is of questionable taxonomic validity leaving two annuals and one perennial species in this subsection with red clover. This subsection is described as containing both perennial and annual species. The calyces are 10 nerved with teeth as long as or shorter than the tube and with an open throat sometimes closed by hairs. Corollas which are usually much longer than the calyces persist in fruit.

Subsection *Intermedia* (Gib. & Belli) Bobrov. contains six species which are apparently also related to red clover. These are: *T. medium* L., *T. pignantii* Brogn. & Bory, *T. heldreichianum* (Gib. & Belli) Hausskn., *T. patulum* Tausch, *T. velebiticum* Deg., and *T. wettsteinii* Dorfl. & Hay. All species in this subsection are perennial with 10 (rarely 12–20) nerved usually glabrous calyces. Throats of calyces generally are closed by callosities. Corollas are much longer than the calyces and usually persist in fruit.

The only other subsection which seems to contain species closely related to red clover is subsection *Alpestria* Gib. & Belli which contains the two perennial species *T. rubens* L. and *T. alpestre* L.. These perennial species generally have 20 nerved calyces that are not completely closed at the throat. Standard petals are as long as the wing petals and corollas persist in fruit.

The possible relationship of red clover to the other species in these three subsections is based on interspecific sexual cross-compatibility and chromosome pairing affinity in the hybrids. Although red clover has been hybridized successfully with species in subsection *Trifolium* and the two related subsections, it is the only species with n=7 chromosomes in any of these three subsections. All other species are n=8. This fact leads to some difficulty in understanding the phylogenetic relationship of red clover to these other species.

2. Hybridization with red clover

Subsection *Stellata* Gib & Belli contain three annual species, *T. stellatum* L., *T. incarnatum* L., and *T. sylvaticum* Gérard ex Loisel. all with 2n=14; however, no successful interspecific hybrids between red clover and species in this group have been reported despite numerous attempts (Taylor et al., 1963; Evans & Denward, 1955). Thus red clover appears to be more closely related to the n=8 species in subsection *Trifolium* than to species in any other subsections. Red clover does appear to be somewhat related to *T. hirtum* All. (2n=10), a member of subsection *Lappacea* Gib. & Belli, based on the production of shriveled seed in the cross of 2x red clover × 2x *T. hirtum* (Taylor et al., 1963). This cross was repeated using 4x red clover × 2x *T. hirtum* and yielded one hybrid plant with gnarled leaves, thick trichomes, and multiple high ploidy level cells (Schwer and Cleveland, 1972b). Crosses with two other species in this section, *T. cherleri* L.(2n=10) and *T. lappaceum* L. (2n=16) were not successful (Quesenberry, 1975).

The two species of subsection *Trifolium* with which red clover has been successfully hybridized are *T. diffusum* (Taylor et al., 1963) and *T. pallidum* (Armstrong and Cleveland, 1970), both annuals. The hybrid of red clover × *T. diffusum* was first made at the diploid level where it was sterile. When both parents were doubled, the amphidiploid of 4x red clover × 4x *T. diffusum* was mostly fertile with an average of 89% pollen stainability. The hybrid was

Fig. 2. 1. *T. pratense* × *T. diffusum* hybrid and parents: L. *T. pratense*, C. Hybrid, R. *T. diffusum*.

generally intermediate between the parents in floral characteristics, but often was less vigorous with reduced vegetative parts compared to either parent, and had yellowish green foliage. The overall growth habit was intermediate between the branching annual type growth of *T. diffusum* and the upright stem crown type growth of red clover, and tended to be annual (Fig. 2.1).

At the diploid level hybrid plants had the expected 2n=15 chromosomes. Analysis of metaphase I chromosome associations showed a high frequency of multivalents with complex configurations of up to 15 chromosomes (Schwer and Cleveland, 1972a). The mean frequency of bivalents was 1.30 per cell and these were always rod shaped suggesting only partial homology of the chromosome arms involved. Anaphase I chromosome disjunction was 8/7 in 67% of the cells studied, but numerous other unbalanced distributions were observed. An average of about 12% of cells at AI and AII had bridges and fragments. The presence of these complex chromosome associations in the diploid hybrid suggests that multiple chromosome rearrangements have occurred between these two species. Thus homologous chromosome segments between species are relatively short leading to various multivalent chromosome configurations. Nevertheless, the low frequency of univalents (1.59) suggests that almost every red clover chromosome has homologous segments with *T. diffusum* chromosomes.

Meiotic analysis of the 4x amphidiploid of 4x red clover × 4x *T. diffusum* (2n=30) supported the above conclusions (Schwer and Cleveland, 1972b). The amphidiploid had a relatively high frequency (9.01) of closed ring bivalents suggesting homogenetic chromosome pairing. Nevertheless, this leaves 12 chromosomes on average associated in some more complex fashion. This amount of allosyndetic chromosome association supports the conclusion that these two species differ by multiple rearrangements of large chromosome segments.

The hybrid of red clover × *T. pallidum* first was attempted at the diploid level and resulted in only shriveled seed (Taylor et al., 1963). Later, a cross of 4x red clover × 2x *T. pallidum* yielded hybrid plants (Armstrong and Cleveland, 1970). Similar to the red clover × *T. diffusum* hybrid, this cross demonstrated reciprocal differences in cross compatibility, and was only successful using 4x red clover as the female. No 4x *T. pallidum* was available and this cross has apparently not been attempted since this report. All the hybrids had the expected 2n=22 chromosomes. In this triploid hybrid 60% of the cells show 7II + 8I as would be expected with homogenetic pairing of the red clover chromosomes. Nevertheless, 36% of the cells had one or sometimes two trivalents. A low frequency of cells also was observed with 8II. These observations and occasional cells with multivalent configurations, also suggest that chromosomes of red clover and *T. pallidum* differ by a series of complex rearrangements. It is unfortunate that the amphidiploid of this hybrid has not been produced as it should yield additional information on the degree of chromosome homology between these two species.

Attempts were made to cross 2x and 4x red clover with 2x *T. noricum* (Quesenberry, 1975), but no plump or shriveled seeds were produced from this cross. This suggests that although morphological features of *T. noricum* place it in subsection *Trifolium,* hybridization barriers between red clover and this perennial species are strong. No 4x *T. noricum* was available for crossing in this study and the possibility remains that an amphidiploid of these two species would be successful. Until the taxonomic validity of *T. mazanderanicum* can be determined through collection of germplasm and determination of chromosome number, any possible role of this putative taxon in the phylogeny of red clover remains speculative.

Red clover appears to be more distantly related to species in section *Intermedia*. Various researchers (Evans and Denward, 1955; Newton et al., 1970; Anderson and Taylor, 1974; Quesenberry, 1975) have attempted to cross red clover with *T. medium* by conventional means without success. Evans (1962a) employed embryo rescue tissue culture on *T. pratense* × *T. medium* hybrids, but was unsuccessful in rearing plants to maturity.

The improvement of tissue culture embryo rescue methods (Collins et al., 1983) have led to the successful production of hybrids of red clover with *T. sarosiense Hazsl.* (Phillips et al., 1982) and with *T. medium* (Merker 1984; Sawai et al., 1990). Hybridization of *T. sarosiense* with 2x red clover

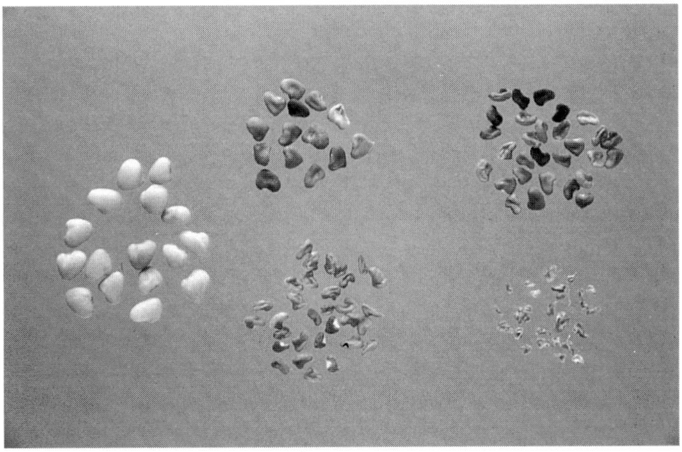

Fig. 2. 2. Seeds dissected from pods of crosses of *T. sarosiense* × *T. pratense*: L. Class I seeds (selfs), Top L. to R. Class II and Class III, Bottom L. to R. Class IV and unfertilized seed coats.

was shown to be dependent on both female and male clones used as parents, although female clones appeared to have larger effects (Taylor et al., 1983). These conclusions were based on the presence of variable size classes of shriveled seeds produced from interspecific matings, suggesting that hybrid embryos from certain combinations developed long before degeneration (Fig. 2.2). No hybrid plants were obtained by normal methods from these shriveled seeds.

Based on the above results, however, heart shaped embryos from superior clonal combinations were excised at 10 to 15 days after pollination and placed onto a newly developed tissue culture medium. Using optimum culture conditions, more than 10% of developing *T. sarosiense* × 2x red clover embryos were rescued *in vitro* (Collins et al., 1983). Plants from five F_1 families were produced and all plants had the expected 2n=31 chromosomes, 24 from *T. sarosiense* and 7 from red clover (Fig. 2.3). Development of flowers on most plants was abnormal and pollen mother cells rarely were observed (Phillips, et al., 1982). Most pistils had only rudimentary style development, but one family exhibited nearly normal, although sterile, floral development. Nevertheless, attempts to backcross the hybrid to red clover were also unsuccessful. The chromosome number of this hybrid was later doubled using an *in vitro* colchicine chromosome doubling method (Anderson et al., 1991); however, the doubled hybrid also showed no pollen development and a scarcity of pollen mother cells. Thus, no analysis of the homology of red clover chromosomes with *T. sarosiense* has been determined. Even so, this lack of development of pollen mother cells indicates a rather major genic incompatibility between these species. That is, the genetic makeup of the two genomes is different enough that gene products fail to function in enough of a

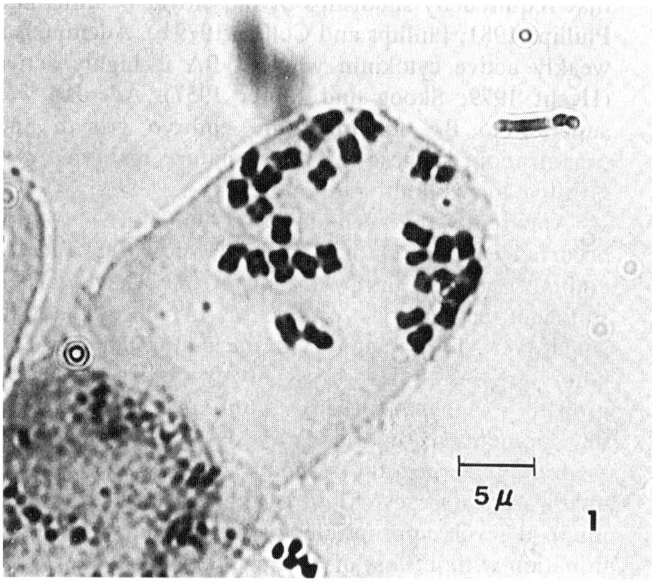

Fig. 2. 3. Chromosomes of the hybrid *T. sarosiense* × *T. pratense* (2n=31).

coordinated manner as to regulate normal microspore development. It would then appear that the chromosomes of red clover have evolved rather far from those of *T. sarosiense*.

Indications of interspecific compatibility between 2x red clover and *T. medium* were reported by Newton et al. (1970). This work using stylar heat treatment of the female parent resulted in a large number of shriveled seeds from apparent post-zygotic embryo abortion. These workers suggested that use of 4x red clover might yield viable hybrids. Later, *T. medium* was crossed with 4x red clover and reciprocally using heat treatment of female parents (Anderson and Taylor, 1974). These workers also found a number of shriveled seed and suggested that post-zygotic embryo abortion was functioning in this cross. Preliminary work at the University of Kentucky on embryo rescue of hybrids from this cross also was not successful.

Research in Sweden (Merker, 1984) using the embryo rescue methods developed by Phillips et al. (1982) was successful in producing mature hybrid plants of 2n=80 *T. medium* × 4x red clover. In this work 2,637 pollinated florets yielded 363 developing embryos of various sizes when excised at 12–14 days after pollination. From these, three hybrid plants were raised. No data on fertility or meiotic analysis of these hybrids was reported. Research in Japan (Sawai and Ueda, 1987) showed that interspecific embryos of *T. medium* × 4x red clover developed somewhat slower than intraspecific embryos, but were relatively normal up to about 10–12 days. However, by 15 days the embryos became vacuolated and finally disintegrated. Hybrid endosperm was also disintegrating by this time. Further research showed that the *T. medium* genotype affected the formation of embryos more strongly than the *T. pratense* genotype (Sawai et al., 1990). Using embryo culture these workers successfully grew nine hybrids to maturity from 217 immature embryos placed into culture. The hybrids came from a 2n=72 line of *T. medium* and had the expected number of 2n=50 chromosomes. No report of meiotic analysis was given, but some plants had enough female fertility that a successful backcross to red clover was reported. At least one of the backcross plants developed rhizomes.

Neither the *T. sarosiense* × red clover nor the *T. medium* × red clover hybrids have provided direct information on chromosomal homology of red clover to these species. However, a review of results indicates that interspecific compatibility is affected by individual clone genotype. The *T. medium* - *T. sarosiense* taxa comprise a complex chromosomal group with numbers ranging from 2n=48 to 2n=80 (Quesenberry and Taylor, 1978, Taylor, 1995). Representatives of these species have been collected from the British Isles across Europe to western Russia and south through the Balkan Peninsula to Turkey, Iran, and Transcaucasia (Zohary and Heller, 1984). The hexaploid *T. sarosiense* may be more limited in distribution to northern and central Europe. It may be that in some regions, wild populations of either *T. sarosiense* or *T. medium* are less evolved from red clover than in others. On a recent germplasm collection expedition in western Bulgaria (Pederson and Quesenberry, 1994), *T. medium* was found frequently growing intermixed with red clover at numerous sites above 1000 meters in the western mountains. Most of these sites also contained *T. alpestre* and to a lesser extent *T. heldreichianum*. Such sympatric occurrence of species often results in the development of strong species isolation barriers (Allaby, 1992) such as found between these species. Although these two perennial species offer potential for incorporation of desirable traits into red clover, additional research will be needed to bring this promise to fruition.

Red clover also appears to be distantly related to species in subsection *Alpestria*. Evans (1962a) obtained apparent hybrid embryos from 2x red clover × *T. alpestre* and 2x red clover × *T. rubens* crosses, but was unsuccessful in obtaining differentiation of these embryos into mature plants. After the development of improved embryo rescue techniques, additional attempts to hybridize 4x *T. alpestre* with 4x red clover were attempted in Sweden (Merker, 1988). In this research about 10% of the pollinated florets yielded hybrid embryos. However, over 60% of the embryos died on the first medium of the embryo culture protocol. Eventually, eight vigorous amphidiploid genotypes were grown to maturity and flowering. Seven of the eight plants had an unexpected 2n=31 chromosomes while only one had the expected 2n=30. They produced normal florets but no pollen. No attempt at meiotic analysis was reported. This hybrid also was reported from research at the University of Kentucky (Phillips et al., 1992). These researchers found only 3 and 0.3% hybrid embryos (experiment 1 and experiment 2) from crosses of

2x *T. alpestre* × 2x and 4x red clover. This low level of embryo production is in fact higher than that of Merker (1988) who reported no hybrids from crosses of 2x *T. alpestre* × 2x red clover. Phillips et al. (1992) were successful in raising one diploid hybrid to maturity. This hybrid had the expected 2n=15 chromosomes, but proved to be both male and female sterile. Unfortunately, no meiotic analysis of this hybrid was reported.

3. Hybridizations among species related to red clover

T. medium (2n=72) has been hybridized with *T. sarosiense.* (2n=48) (Quesenberry and Taylor, 1978). Hybrid plants from the reciprocal cross of *T. sarosiense* × 2n=72 *T. medium* all showed various types of chlorosis indicating genic-cytoplasmic incompatibility between the two species. Meiotic analyses of several different individual *T. medium* × *T. sarosiense* hybrids always showed greater than 24 II (range of 25.96 to 27.66) suggesting that some *T. medium* chromosomes were pairing autosyndetically in addition to the 24 *T. sarosiense* chromosomes pairing with 24 *T.medium* chromosomes. All hybrids had relatively high pollen fertility (mean of 89%). These findings suggest that these two taxa differ primarily by ploidy level increase followed by small genetic and cytoplasmic changes.

Magulaev (1989) reported that among *T. medium* accessions from the Crimea and Caucasus over 30% had 2n=56 chromosomes, exactly intermediate between 2n=48 *T. sarosiense* and a 2n=64 *T. medium.* In contrast, Taylor (1995) reported no intermediate chromosome numbers between 2n=48 and 2n=64 among 35 accessions of *T. sarosiense* and *T. medium* in the United States germplasm collection. Nevertheless, Taylor (1995) chose to follow the classification of Zohary and Heller (1984) who classify *T. sarosiense* as *T. medium* Var. *sarosiense* (Hazsl.) Savul & and Rayss. The diversity of findings in the above reports does not resolve conclusively the question of species rank for *T. sarosiense.* Although some minor crossing barriers of hybrid chlorosis exist between the two species (Quesenberry and Taylor, 1978), these two taxa appear to be closely related. Both taxa contain traits that would be desirable to intergress into red clover, such as rhizomatous root habit and disease resistance.

Earlier work had demonstrated that the two species of subsection *Alpestria*, *T. alpestre*, and *T.rubens* could be hybridized at the diploid level yielding a partially fertile F_1 (Quesenberry and Taylor, 1978). Meiotic analysis of this hybrid at the diploid level showed a mean of 7.77 II at MI with very few laggards or micronuclei at second division. This same research also showed that *T. alpestre* and *T. heldreichianum*, a member of subsection *Intermedia* could be hybridized readily. This hybrid also showed a high degree of chromosome homology (mean of 7.95 II) between these two species. In addition hybrids of *T. rubens* with *T. noricum* were all pale green and did not survive past the first trifoliate leaf stage. This hybrid suggested that these species had potential for hybridization with red clover, although no hybrids of these species with red clover were produced by conventional means. In related research, *T. sarosiense* and 4x *T. alpestre* were hybridized successfully (Quesenberry and Taylor, 1978). Meiotic analysis of this hybrid also showed a high degree of chromosome homology between *T. alpestre* and *T. sarosiense*, with a mean of 13.86 II. Although varied interpretations of this data are possible, the presence of occasional multivalents suggests that the 14 *T. alpestre* chromosomes paired with 14 *T. sarosiense* chromosomes. The hybrid had a mean pollen stainability of 63%. These results show that species of subsection *Alpestria* have strong chromosome homology with the species of subsection *Intermedia*.

Over the years, numerous researchers have attempted to cross red clover with various *Trifolium* species in other subsections and sections of the genus (Evans and Denward, 1955; Kazimierski et al., 1972; Quesenberry, 1975) without success. Perhaps one of the most frequent unsuccessful attempts has been the cross of red clover with alsike clover (*T. hybridum* L.). Since this species once was considered a hybrid of red clover and white clover (*T. repens* L.), many researchers have attempted this cross. Hyrkas-Lyytikainen and Kivinen, 1989 have reported the application of embryo rescue culture techniques to this cross without success. These researchers did report development of globular shaped embryos, but all died on the culture medium. Although these embryos may have been of hybrid origin, their failure where embryo culture of crosses with species in subsection *Intermedia* were successful, illustrates the strong post-fertilization barrier between red clover and alsike clover or white clover. Starzycki (1959) reported hybridization between red clover and white clover which was never verified. Recently, a hybrid of red clover with white clover has been reported as the result of embryo rescue (Repkova et al., 1991). A discussion of the tissue culture medium used to produce this hybrid is included in Chapter 14. These workers report obtaining 104 hybrid plants from crosses of *T. repens* f. *hollandricum* × both diploid and tetraploid forms of red clover. They report hybrid isozyme patterns intermediate between the parents, but no details on vigor, chromosome number, or fertility of the hybrid.

4. Barriers to interspecific hybridization

Barriers to interspecific hybridization with red clover can generally be grouped into two categories, pre- or post-zygotic (Evans 1962a, b). Attempted methods for overcoming pre-zygotic barriers include stylar temperature pretreatment (Anderson and Taylor 1974), comparison of stylar lengths and pollen tube growth rates (Evans, 1962b), ploidy level manipulations (Taylor et al, 1963; Quesenberry and Taylor, 1978; Merker, 1984), mentor pollen (Taylor et.al.,

1980), and self-compatibility of the female parent. All these efforts support the conclusion that manipulations affecting pollen tube growth and development in the stigma and style are not likely to increase the success of hybridization of related species with red clover.

In general, the only pre-zygotic method which has been successful in improving the frequency of hybrid progeny was manipulation of ploidy levels. Parrott and Smith (1986) proposed that most interspecific hybridization results in sections *Lotoidea* and *Trifolium* could be explained on the basis of the endosperm balance number (EBN) hypothesis (Johnston et al., 1980). They suggested that diploid red clover and *T. diffusum* have an EBN of 3 whereas *T. pallidum* has EBN=6 and *T. alpestre* and *T. rubens* have EBN=4. These workers tested their hypothesis using a line of red clover which produced some unreduced female gametes. They showed that the EBN hypothesis explained the results of crosses between red clover and either 2x or 4x *T. diffusum* and 2x *T. pallidum*. Some of the more recent results of crossing red clover with *T. medium* and *T. alpestre* do not appear to support the EBN hypothesis.

The primary post-zygotic barrier to interspecific hybridization of related species with red clover appeared to be embryo abortion due to endosperm breakdown (Evans, 1962b). Using improved embryo culture media, embryo rescue has successfully produced hybrids of *T. sarosiense,* (Collins et al., 1983), *T. medium* (Merker, 1984; Sawai et al., 1990), and *T. alpestre* (Merker, 1988; Phillips et al., 1992) with red clover. Other post-zygotic hybridization barriers include chlorosis of hybrid seedlings (Taylor et al., 1963; Quesenberry and Taylor, 1976, 1977), and lack of normal development of floral reproductive structures (Phillips et al., 1982). These barriers would be what Dobzhansky et al. (1977) described as hybrid inviability or hybrid weakness. Still other barriers were male or female sterility of the F_1 hybrids (Taylor et al., 1963; Merker, 1984;). Dobzhansky et al. classified the type of barriers where chromosome doubling of sterile hybrids can overcome fertility problems (Taylor et al., 1963) as chromosomal sterility. Isolation barriers seen in other hybrids where fertility is not restored by chromosome doubling (Anderson et al., 1991) have been referred to by Dobzhansky et al. as genic sterility. It appears all of the barriers discussed by Dobzhansky et al. (1977) function in one or more of the hybrids between red clover and the various species with which it will hybridize.

Identification of specific genotypes of related species may improve the frequency of embryo production (Collins et al., 1983), and yield F_1 hybrids with increased fertility (Sawai et al., 1990). Based on these findings, additional effort appears merited to identify genotypes of *T. sarosiense, T. medium,* and *T. alpestre* which may "nick" better with red clover in interspecific crosses.

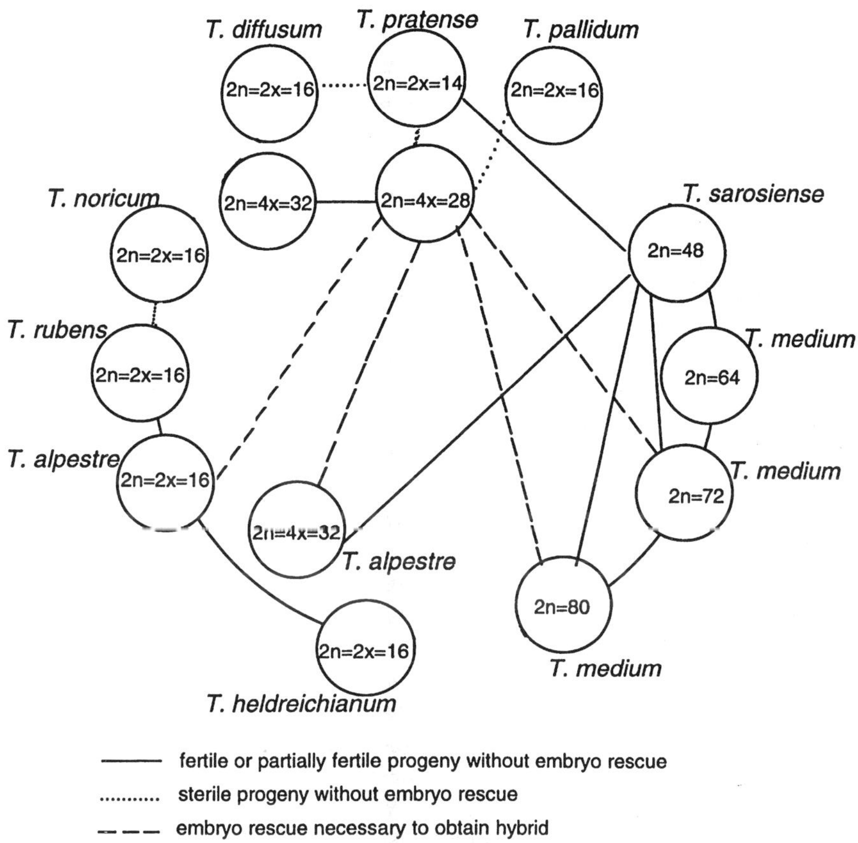

Fig. 2. 4. Crossing relationships among species related to red clover.

5. Species groupings based on interspecific hybridization

Based on the records of attempted and successful interspecific hybrids between related species and red clover, and the ease with which crosses are made, some natural groupings of species are apparent. Fig. 2.4, modified from Cleveland (1985), attempts to summarize these species groups. In this grouping red clover is grouped with the two diploid annual species *T. diffusum* and *T. pallidum*. The polyploid *T. medium-sarosiense* complex forms another group, and the diploid perennial species of *T. alpestre, T. heldreichianum, T. noricum,* and *T. rubens* form a third group.

Harlan and de Wet (1971) and Harlan (1975) proposed the concept of gene pools to provide a genetic perspective on classification of the relatives of cultivated plants. They proposed that the related species of cultivated plants can be classified as members of either the primary, secondary, or tertiary gene pool of the cultivated species. The primary gene pool corresponds to the traditional concept of a biological species. Among members of this pool, crosses are made easily, hybrids are fertile with good chromosome pairing, and transfer of genes is simple. With most cultivated species there are likely to be wild or weedy forms as well as the cultivated races. In red clover this pool would include the six botanical varieties described by Zohary and Heller (1984) and probably the doubtful taxon *T. mazanderanicum*.

The secondary gene pool includes all biological species that will cross with the cultivated species by "normal" means. Gene transfer is possible, but there are likely to be barriers. Hybrids are often semi to nearly sterile and chromosomes often pair poorly. Some hybrids may be weak and difficult to rear to maturity. For red clover, the species *T.diffusum* and *T. pallidum* are perhaps the only two species which fall in this gene pool. It is possible that *T. hirtum* should also be considered a member of this gene pool, but additional crosses with different ploidy levels are needed to resolve this status.

The tertiary gene pool consists of species which can be crossed with the cultivated species only with great difficulty, often including embryo culture, chromosome doubling, or bridge crosses. Hybrids with species in this group are likely to be mostly sterile and gene transfer very difficult. Nevertheless, these species often contain traits desirable for incorporation into the cultivated species. Although no known meiotic analyses of hybrids between *T. sarosiense, T. medium,* or *T. alpestre* with red clover have been published, the crossability of these species with red clover suggests that they fall in the tertiary gene pool of red clover. Since *T. alpestre* crosses readily with *T. heldreichianum* and *T. rubens* by bridge crosses, these species also become members of the tertiary gene pool of red clover. *T. noricum* appears to have some interspecific compatibility with both *T. rubens* and *T. alpestre* and thus also can be considered member of the tertiary gene pool of red clover. Additional research is needed to define the status of the recently collected species *T. velebiticum, T. pignantii,* and *T. patulum*, as well as the uncollected species *T. wettsteinii.*

For red clover, only plant introductions, improved cultivars, and botanical varieties are included in the primary gene pool with red clover. The apparent major chromosomal rearrangements leading to partial sterility in hybrids between red clover and *T diffusum* or *T. pallidum* place these species in the secondary gene pool. All the diploid and polyploid perennial species would appear to fall in the tertiary gene pool. The status of *T. repens* as a component of the red clover tertiary gene pool awaits additional information on the fertility of the reported hybrid. As more research is conducted, particularly in the area of molecular genetics, a clearer definition of the relationship of red clover to these other *Trifolium* species may be possible.

References

Allaby, M. 1992. *The Concise Oxford Dictionary of Botany*. Oxford University Press, Oxford.

Anderson, J.A., C. Mousset-Declas, E.G. Williams and N.L. Taylor. 1991. An *in vitro* chromosome doubling method for clovers (*Trifolium* spp.). *Genome.* 34:1–5.

Anderson, M.K. and N.L. Taylor. 1974. Effect of temperature on intra- and interspecific crosses of diploid and tetraploid red clover, *Trifolium pratense* L. *Theor. Appl. Genet.* 44:73–76.

Armstrong, K.C. and R.W. Cleveland. 1970. Hybrids of *Trifolium pratense* × *Trifolium pallidum*. *Crop Sci.* 10:354–357.

Cleveland, R.W. 1985. Reproductive cycle and cytogenetics. In N.L. Taylor (ed.), *Clover Science and Technology. ASA-CSSA-SSSA Mono.* 25:71–110. Madison, WI.

Collins, G.B., N.L. Taylor and G.C. Phillips. 1983. Successful hybridization of red clover with perennial *Trifolium* species via embryo rescue. In *Proc. Intern. Grass. Cong.* 14:168–170. Westview Press, Boulder, CO.

Dobzhansky, T., F.A. Ayala, G.L. Stebbins and J.W. Valentine. 1977. *Evolution.* W.H. Freeman and Co. San Francisco.

Evans, A.M. 1962a. Species hybridization in *Trifolium* I. methods of overcoming species incompatibility. *Euphytica.* 11:164–176.

Evans, A.M. '62b. Species hybridization in *Trifolium* II. investigating the pre-fertilization barriers to compatibility. *Euphytica.* 11:256–262.

Evans, A.M., and T. Denward. 1955. Grafting and hybridization experiments in the genus *Trifolium. Nature.* 175:687–688.

Harlan, J.R. 1975. *Crops and Man.* American Society of Agronomy, Madison, WI.

Harlan, J.R. and J.M.J. de Wet. 1971. Toward a rational classification of cultivated plants. *Taxon.* 20:509–517.

Hyrkas-Lyytikainen, K. and M. Kivinen. 1989. Attempts to hybridize *Trifolium pratense* and *T. hybridum. Hereditas.* 111:25–29.

Johnston, S.A., T.P.M. den Vijs, S.J. Peloquin and R.E. Hanneman, Jr. 1980. The significance of genic balance to endosperm development in interspecific crosses. *Theor. Appl. Genet.* 57:5–9.

Kazimierski, T., E.M. Kazimerska, and C. Strzyzewska. 1972. Species crossing in the genus *Trifolium* L. *Genetica Pol.* 13:11–32.

Magulaev, A.Y. 1989. Chromosome numbers in *Trifolium medium* in the flora of the Crimea and Caucasus (In Russian). *Phitologiya i Genetika.* 23:26–29.

Merker, A. 1984. Hybrids between *Trifolium medium* and *Trifolium pratense. Hereditas.* 101:267–268.

Merker, A. 1988. Amphidiploids between *Trifolium alpestre* and *Trifolium pratense. Hereditas.* 108:267.

Newton, D.L., W.A. Kendall and N.L. Taylor. 1970. Hybridization of some *Trifolium* species through stylar temperature treatments. *Theor. Appl. Genet.* 40:59–62.

Parrot, W.A., and R.R. Smith. 1986. Evidence for the existence of endosperm balance number in the true clovers (*Trifolium* spp.) *Canad. J. Genet. Cytol.* 28:581–586.

Phillips, G.C., G.B. Collins and N.L. Taylor. 1982. Interspecific hybridization of red clover (*Trifolium pratense* L.) with *T. sarosiense* Hazsl. using *in vitro* embryo rescue. *Theor. Appl. Genet.* 62:17-24.

Phillips, G.C., J.W. Grosser, S. Berger, N.L. Taylor and G.B. Collins. 1992. Interspecific hybridization between red clover and *Trifolium alpestre* using *in vitro* embryo rescue. *Crop Sci.* 32:1113–1115.

Pederson, G.A. and K.H. Quesenberry. 1994. Collection of *Trifolium* germplasm in southwestern Bulgaria. In *Proceedings of the 13th Trifolium Conference.* 22–24 June 1994. Charlottetown, Prince Edward Island, Canada.

Quesenberry, K.H. 1975. Interspecific hybridization of perennial *Trifolium* species related to red clover. Ph.D. dissertation. Univ. of Kentucky, Lexington, KY (Dissertation Abstr. 36/09-B.DCJ76–06143) p. 4256.

Quesenberry, K.H. and N.L. Taylor. 1976. Interspecific hybridization in *Trifolium* L., Sect. *Trifolium* Zoh. I. diploid hybrids among *T. alpestre* L., *T. rubens* L., *T. heldreichianum* Hausskn., and *T. noricum* Wulf. *Crop Sci.* 16:382–386.

Quesenberry, K.H. and N.L. Taylor. 1977. Interspecific hybridization in *Trifolium* L., Sect. *Trifolium* Zoh. II. fertile polyploid hybrids between *T. medium* L., *T. sarosiense* Hazsl. *Crop Sci.* 17:141–145.

Quesenberry, K.H. and N.L. Taylor. 1978. Interspecific hybridization in *Trifolium* L., Sect. *Trifolium* Zoh. III. partially fertile hybrids of *T. sarosiense* Hazsl. × *T. alpestre* L. *Crop Sci.* 18:536–540.

Repkova, J., B. Nedbalkova and J. Holub. 1991. Regeneration of plants from zygotic embryos after interspecific hybridization within the genus *Trifolium* and electrophoretic evaluation of hybrids. OSEVA *Research Institute for Fodder Plants*. 12:7–14. (From Plant Breeding Abstracts 1993).

Sawai, A. and S. Ueda. 1987. Embryo development of the hybrid of *Trifolium medium* L. × *4x Trifolium pratense* L. *J. Japan. Soc. Grassl. Sci.* 33:157–162.

Sawai, A., S. Ueda, M. Gau and K. Uchiyama. 1990. Interspecific hybrids of *Trifolium medium* L. × 4x *Trifolium pratense* L. obtained through embryo culture. *J. Japan. Soc. Grassl. Sci.* 33:157-162.

Schwer, J.F. and R.W. Cleveland. 1972a. Diploid interspecific hybrids of *Trifolium pratense* L., *T. diffusum* Ehrh., and some related species. *Crop Sci.* 12:321–324.

Schwer, J.F. and R.W. Cleveland. 1972b. Tetraploid and triploid interspecific hybrids of *Trifolium pratense* L., *T. diffusum* Ehrh., and some related species. *Crop Sci.* 12:419–422.

Starzycki, S. 1959. Zmienosc miezancow koniczyny bialej (*Trifolium repens* L.) Z koniczyna czerwona *Trifolium pratense* L. *Nasiennictwo*. 3:277–319. (In *Pl. Breed. Abstr.* 30:546–547, 1960).

Taylor, N.L. 1995. Characterization of the United States germplasm collection of zigzag clover (*Trifolium medium* L.) *Genet. Resour. & Crop Evol.* 42:43–47.

Taylor, N.L., G.B. Collins, P.L. Cornelius and J. Pitcock. 1983. Differential interspecific compatibilities among genotypes of *Trifolium sarosiense* and *T. pratense*. In *Proc. Intern. Grassl. Congr.* 14:165–168. Westview Press, Boulder, CO.

Taylor, N.L., R.F. Quarles and M.K. Anderson. 1980. Methods of overcoming interspecific barriers in *Trifolium*. *Euphytica*. 29:441–450.

Taylor, N.L., W.H. Stroube, G.B. Collins and W.A. Kendall. 1963. Interspecific hybridization of red clover (*Trifolium pratense* L.). *Crop Sci.* 3:549–552.

Zohary, M. and D. Heller. 1984. *The Genus Trifolium*. The Israel Academy of Sciences and Humanities. 606 pp.

3. Reproductive Biology, Genetics and Evolution

Gametogenesis and the reproductive cycle of red clover were reviewed by Cleveland (1985); however, a brief review is presented here to facilitate development of the remainder of this topic. This review relies heavily on the work of Hindmarsh (1964) and Mackiewicz (1965). Reproductive topics covered include: 1) anther and pollen development, 2) ovule and embryo sac development, 3) fertilization and pollen tube growth, 4) self incompatibility, and 5) embryo development. A brief review of cytological methods used with red clover is also presented.

1. Cytological methods

Methods for obtaining satisfactory preparations of red clover tissues for study of reproductive biology have changed as new techniques have been introduced. Some of the earliest reports of chromosome counts in red clover were based on sectioned tissue, but most recent reports have involved use of fixed and squashed materials. As with most other species, workers with red clover generally have not reported comparisons of various methods, but rather only the techniques which were used in their research. The two major studies of anther and microspore and of ovule development that are reviewed in this chapter used tissue that was fixed in either formalin, acetic acid, alcohol (FAA) or Navaschin's fixative, imbedded in paraffin, and sectioned at 8 to 14 microns (Hindmarsh, 1964; Mackiewicz, 1965). These methods appear to be satisfactory for anatomical study of sporogenesis. More recent reports have used a 18:1:1 mixture of 50% ethanol : 37% formaldehyde : acetic acid for 2 h followed by paraffin embedding (Sawai and Ueda, 1987). Sections were usually stained in hematoxylin and often with a counter stain of fast green. Ovule clearing techniques have not been used in red clover.

Somatic chromosome counts of red clover have been obtained by various protocols. The most frequently encountered procedure uses actively growing root tips, often from germinating seeds or newly rooted vegetative cuttings. Use of either of these tissues will allow the investigator to collect root tips from a "clean" substrate such as filter paper or water, avoiding the problems of sand and silt contamination of the specimen. A pretreating agent is frequently used to increase the mitotic index in the tissue and improve visibility of the small red clover chromosomes. The method of Anderson and Taylor (1974) involves pretreatment for 4 to 5 h in 0.003 M 8-hydroxyquinoline followed by fixation in a 3:1 (v:v) solution of 95% ethanol:glacial acetic acid. Some investigators have added 2 drops of a maltose solution to the pretreatment solution (Taylor et al., 1979). Tissue which is to be used within a few days can remain in this fixative, otherwise it should be transferred to 70% ethanol for long term storage. Root tips are then hydrolyzed in 1N Hcl at 60 C for 6 to 10 min. An alternative to this hydrolysis procedure is use of 5N HCl at room temperature for 15 min (Sawai et al., 1990). Following hydrolysis, root tips are rinsed in water, stained in Feulgen for variable time periods (10–30 min), and softened for up to 2 h in 45% acetic acid. Root tips are then placed in a small drop of the 45% acetic acid where the root cap is carefully removed. The meristematic region of the tip is then excised and the remainder of the root is discarded. The meristematic tip is then squashed in 1% acetocarmine. Propionic-carmine or propionic-orcein may also be used as the final mounting stain (Schwer and Cleveland, 1972a). For observation of meiotic chromosomes in developing anthers, the method of Quesenberry and Taylor (1976) is often used. Developing flower buds are collected between 0930 and 1030 h when the calyx teeth are barely visible between the two stipules of the uppermost leaves. These surrounding stipules are then removed and the heads are then fixed in a solution of 95% ethanol : chloroform : glacial acetic acid (6:3:1/v:v:v) for 12 to 24 h. Heads may remain in this solution for 3 to 5 d, but for longer term storage, should be transferred to 70% ethanol. The entire head is then removed and stained in Snows alcoholic acid carmine for 24 h (Snow, 1963). The head is removed from the stain, rinsed in 70% ethanol, and florets are removed individually and arranged from top to bottom of the head to aid in identification of where the various stages of meiosis may be located on the head. The 10 anthers are then dissected from a given floret on a clean microscope slide. Once all extraneous debris is removed, a drop of 1% acetocarmine is added and the anthers are gently squashed to observe the stages of meiosis. Others have used anthers softened in 1N HCl at 65C for 5 to 15 sec and squashed in propionic-orcein to observe meiotic chromosomes of red clover (Schwer and Cleveland, 1972a). Another method uses 6:3:2 (v:v:v) methanol : chloroform : saturated iron acetate in propionic acid, followed by staining according to the method of Snow (1963) and squashing in modified carbol fuchsin with mounting in Hoyer's medium (Taylor and Wiseman, 1987). A similar method was used by Parrot and Smith (1984) to study the cytology of plants with a high frequency of 2n pollen production.

Various methods have been used to access pollen quality. Perhaps the simplest is to dust dry pollen onto a clean microscope slide, add a small drop of 1% acetocarmine, cover gently with a cover slip and determine the percentage of pollen grains which have well stained cytoplasm (Quesenberry and Taylor,

1976). With some genetic aberrations, variation in pollen size and staining may be observed creating a need to obtain additional information about pollen quality. *In vitro* pollen germination may be used in this circumstance. For optimal red clover pollen tube germination and elongation an *in vivo* medium containing 25% (w:v) of sucrose, 50 mg/l boric acid, and 100 mg/l calcium carbonate solidified with agar at 12.5/g has been used (Kendall and Taylor, 1965).

2. Anther and microspore development

The undifferentiated red clover anther is ovid in cross section, but as it undergoes cell division it becomes nearly rectangular (Hindmarsh, 1964), . With further division five to six rows of archesporial cells may be distinguished. These cells differentiate into endothecium, middle layer, tapetum, and microsporocytes (tissue layers under the epidermis from outer to inner, respectively). By the time the microspore mother cell is ready to begin meiosis, the endothecium and middle layers become vacuolated and flattened against the epidermis. During this developmental period the two lobes of the anther become clearly defined. Vascular connections of each lobe to the filament also become clearly defined. Tapetal cells increase in size just prior to meiotic division. After the tetrads are produced, the inner wall of each tapetal cell breaks down but the cell contents remain *in situ* where they are absorbed. As the anther matures, the cells of the endothecium become highly vacuolated with fibrous wall thickenings.

The microspore mother cells become distinctly rounded and enlarged prior to the onset of meiosis. The meiotic process follows a normal pattern in diploid red clover. Chromosomes of red clover are difficult to distinguish in leptotene or zygotene of prophase I, appearing as lightly stained strands. Although distinct structures can be observed at pachytene, detailed analyses which would reveal inversion or translocation heterozygosity generally are not possible in red clover (author's personal observation). However, by diplotene or diakinesis, clearly distinguishable bivalent chromosome pairs suitable for chiasma analyses are present with one pair of chromosomes associated with the nucleolus.

By the beginning of metaphase I (MI) the nucleolus disappears and the chromosomes form distinct bivalents. In normal diploid red clover the bivalents are consistently ring shaped as would be expected from metacentric or submetacentric chromosomes with crossovers in each arm. The presence of rod bivalents in the *T. pratense* × *T. diffusum* hybrid is evidence that only short homologous segments between chromosome arms of these two species are similar (Schwer and Cleveland, 1972a). No studies of meiosis in hybrids between the various taxonomic varieties of red clover have been reported which would allow comparison of types of bivalent formation between more distantly related forms of red clover.

Anaphase I (AI) segregation shows seven chromosomes moving to each pole. At telophase I (TI) the chromosomes relax back into a diffuse nucleus. As is typical for legumes no cytokinesis occurs between first and second division of meiosis. Metaphase II (MII) shows two distinct groups of seven chromosomes with their axes oriented at right angles. As will be discussed later, mutant genotypes may be identified in which the two axes are parallel leading to production of first division restitution (FDR) 2n gametes. In normal diploid red clover four groups of seven chromatids can be distinguished moving to the poles at anaphase II. No reports of AI or AII bridges, indicating paracentric inversion heterozygotes are known in normal red clover. Evidence of up to three such inversions were observed in the *T. pratense* × *T. diffusum* hybrid (Schwer and Cleveland, 1972a). Quadravalent rings resulting from reciprocal translocations between lines of red clover were not found.

After telophase II (TII), simultaneous quadrapartitioning of the cytoplasm by furrowing and vacuolation results in tetrad microspores. The tetrad products of microgametogenesis are generally tetrahedral, but may be decussate or isobilateral (Hindmarsh, 1964). Each microspore forms its own membrane while the four TII nuclei are still enclosed in the original microspore mother cell wall which becomes thickened and gelatinous. Tetrads appear to be released by digestion of this enclosing wall.

The free microspores are non-vacuolate and often angular, but with development become elongated oval and keg shaped. First division of pollen mitosis occurs during this time resulting in formation of a densely staining generative nucleus and a more diffuse and lightly staining tube nucleus. However, pollen mitosis is difficult to observe in collected red clover anthers. After first division of pollen mitosis the exine wall of the pollen grain thickens except over three elongate areas which persist as germ pores. Pollen is shed as binucleate with division of the generative nucleus into the two male gametes occurring in the pollen tube.

The septum between the two lobes of the anther breaks down prior to dehiscence (Hindmarsh, 1964). This results in a common cavity between the two lobes. Dehiscence usually occurs by rupture along the longitudinal line between the two anther halves. This commonly occurs before the flower reaches full elongation, but the dehised pollen is held in place in the fused keel petal until tripped. Fig. 3.1 shows the stigma and mature anthers of red clover just prior to anther dehiscence.

A distinct difference between diploid and tetraploid red clover is the shape of mature pollen (Taylor et al., 1976). When mounted dry and observed at low power under a microscope, tetraploid pollen is angular to tetrahedral in shape and larger in size than diploid pollen which is oblong oval shaped. Pollen of triploids was a mixture of oblong-oval and tetrahedral shapes varying somewhat in size (Taylor and Wiseman, 1987). Although wet mounted tetraploid pollen was 16% larger than diploid (Mackiewicz, 1965), this method is not as useful for distinguishing ploidy differences as the dry mounted method.

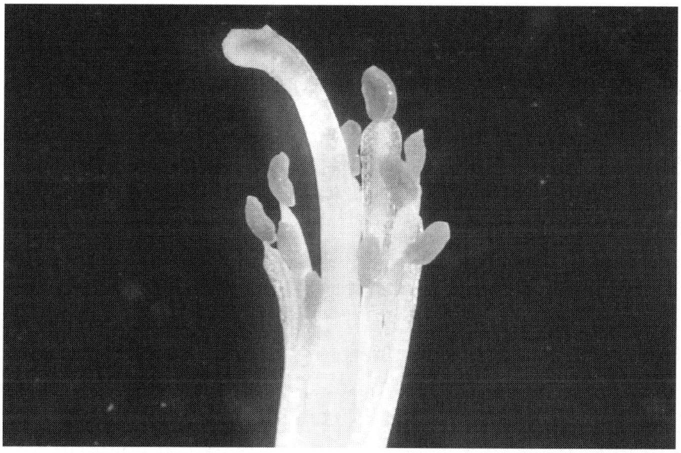

Fig. 3. 1. Red clover stigma and mature anthers just prior to pollen dehiscence.

The difference in tetraploid pollen size and shape allows for rapid screening of plants for induction of tetraploidy (Taylor et al., 1976) or identification of 2n gamete producing plants (Parrott and Smith, 1984) (See chapter 13 on tetraploid red clover for additional details). Using this method to identify 2n pollen producers, three cycles of recurrent selection increased frequency of 2n pollen producing individuals from 0.04% of the population to 47.4%. Heritability of 2n pollen formation was estimated at 0.50 (Parrott and Smith, 1986b).

The potential advantages of using gametic non-reduction for producing tetraploid red clover are discussed in detail in chapter 13. Maximum heterozygosity in unreduced gamete derived populations is obtained if the 2n gametes are derived via meiotic abnormalities leading to FDR gametes as opposed to second division restitution (SDR) gametes. On average FDR gametes transfer about 80% of parental heterozygosity to offspring compared to only about 40% for SDR gametes (Hermsen, 1984). Various cytological abnormalities may lead to the production of 2n gametes in red clover, but regardless of origin there are only two genetic consequences, FDR or SDR (Mendiburu and Peloquin, 1976; Mok and Peloquin, 1975). The type of 2n pollen producing mutant identified in red clover was parallel spindles at second division of meiosis (Parrott and Smith, 1984) which has the genetic consequence of SDR type gametes. Red clover 2n gamete production decreased as temperatures increased from 12 to 25 C (Mousset-DeClas et al., 1992). In this work both intragenotypic and intergenotypic variation also were significant sources of variation in the amount of 2n pollen produced.

3. Ovule and embryo sac development

The red clover ovary regularly contains two ovules of which only one usually develops into a seed. However, in some high seed-yielding plants both ovules develop mature embryo sacs and often mature two seeds. Thus a review of embryo sac development appears merited. This discussion primarily follows Hindmarsh (1964).

The primordia for integument development first appear as folds at the base of the nucellus, and as the result of differential growth the ovule becomes half curved or campylotropous in form. Both outer and inner integuments begin to develop but the outer rapidly overgrow the inner. The integuments continue to grow during the time of megasporogenesis and megagametogenesis. By the time the embryo sac is mature, the outer integument has completely enclosed it leaving only a micropylar opening. At this time the epidermal cells of the outer integument became columnar and continuous with similar cells of the funicle.

A single (occasionally two) archesporial cell was formed simultaneously with the emergence of the integument primordia. This cell divided to form an outer parietal cell and the inner primary sporogenous cell. The outer parietal cell divided several times forming a two to three cell layer over the megaspore mother cell. In contrast to microsporogenesis, each division of meiosis was followed by cytokinesis resulting in a linear tetrad of megaspores. However, the chalazal megaspore undergoes a rapid size increase while the remaining three break down.

The remaining megaspore undergoes a further increase in size and the nucleus divides. These two daughter nuclei move to the ends of the embryo sac where a second nuclear division results in a four-nucleate embryo sac. A third division yields the eight-nucleate state with two groups of four nuclei at each end of the embryo sac. By further development one nucleus from each end merges to form the binucleate endosperm cell. Prior to fertilization this endosperm cell enlarges over the egg cell in a hood-like manner.

The reason for failure of one of the ovules to develop a mature embryo sac or for failure of the second embryo sac to begin developing an embryo are not well understood. Nevertheless, low seed-yielding plants usually have a lower percentage of ovaries with at least one normal embryo sac, have a lower percentage of ovaries showing developing embryos and endosperm 48 h after pollination, and ultimately have a lower percentage of florets that set at least one seed.

4. Pollen tube growth and self-incompatibility

The stigma of red clover is receptive of pollen by or before the time the flower petals have reached full extension. The style of red clover forms a slightly swollen entasis below the stigma and this open cavity contains a watery secretion (Heslop-Harrison and Heslop-Harrison, 1982). Red clover pollen tube growth is governed by the gametophytic self-incompatibility system. Nevertheless, intraspecific pollen of either compatible or incompatible genotypes germinates readily on the stigma. After a short period of rapid growth, tubes of incompatible pollen types are retarded and finally stop growth. Most incompatible tubes stop growth in the entasis region of the style before reaching the lower, more narrow, region of the style. Mean style length of diploid red clover is about 9.25 mm, and the maximum pollen tube growth rate for red clover is about 2.0–2.5 mm/hr (Evans, 1962). Therefore, red clover pollen tubes should be approaching the micropyle in 5–6 h after pollen germination. Circumstantial evidence of polyploid induction from N_2O treatment of pollinated red clover (Taylor et al., 1976) suggests that first division of the fertilized zygote takes place between 24 and 48 h after pollination. At 24 h after pollination, seven of eight embryos in fertilized embryo sacs still were undivided but the endosperm was three to eight nucleate (Pandey, 1955). At 48 h, the fertilized embryos were two to six celled and endosperm was seven to 19 nucleate. In Russia, the optimum time for pollination of red clover was from 1000 to 1500 h at an air temperature of 21 to 24 C and relative humidity of 40–50% (Bekuzarova et al., 1987). This work also showed evidence of selective fertilization which depended not only on the parental genotypes, but also on the developmental age of the androecium and gynoecium and the quality and quantity of pollen.

Although normal diploid red clover exhibits typical gametophytic self-incompatibility a limited amount of pseudo-self compatibility occurs in most plants and by this method homozygous S alleles individuals (S_1S_1, S_2S_2) can be produced from normal heterozygous (S_1S_2) parents (Taylor, 1982). Pseudo-self-compatibility may be increased by exposing flowers to temperatures between 30 and 40 C during development (Leffel, 1963; Kendall and Taylor, 1969). The number of different S alleles was shown to be extensive by Williams and Williams (1947) who identified at least 41 different alleles from a possible 48 in 24 individual red clover plants. Red clover apparently generates new S specificities (Denward, 1963; Anderson et al., 1974; Taylor, 1982). New S alleles may result from crossing over in the S locus which was found to be tightly linked to flower color (Denward, 1963). Origin of new S alleles also has been attributed to inbreeding (Anderson et al., 1974), and to intracistronic recombination (de Nettancourt, 1977). In a synthesized double cross hybrid, the four expected S allele genotypes were identified, but several other unexpected genotypes were also found. These new S alleles may have arisen by intracistronic recombination or somatic crossing over (Taylor, 1982).

Self-compatible plants of red clover have been identified by various workers. The genetic factor controlling expression of this trait is allelic with the self-incompatibility alleles and is designated S_f. In crosses S_f was dominant to S_x. Plants assumed to be heterozygous S_fS_x when test crossed to unrelated self-incompatible plants (e.g. S_xS_y) segregate approximately 1:1 for self-fertility and self-incompatibility as expected. Although inbreeding depression is normally severe in red clover, stocks of the S_f allele have been maintained through selection for several generations and are available for research (Quesenberry et al., 1991).

The physiological factor(s) imparting the self-incompatibility response in red clover is not fully known. Red clover styles are hollow and the hollow canal below the stigma contains a watery secretion through which pollen tubes grow after penetrating the stigma (Heslop-Harrison and Heslop-Harrison, 1982) . The stylar fluid contains glycoproteins which may be involved in the incompatibility response. In petunia (*Petunia inflata*) the so called S proteins are ribonucleases involved in recognition and rejection of self pollen (Lee et al., 1994; Huang et al., 1994). Transgenic plants have been used to show that inhibition of synthesis of S_3 proteins in a S_2S_3 plant through transformation with the antisense S_3 gene results in failure of the transgenic plant to reject S_3 pollen. Transforming a S_1S_2 petunia plant with the normal transgene for S_3 also made the plant reject S_3 pollen. In other work a mutagenized S_3 gene with the codon for His-93 (implicated in ribonuclease activity) was replaced with asparagine. A S_1S_2 plant was then transformed with this gene. Although a normal level of the S_3 protein was detected, no ribonuclease activity was present and the plant did not reject S_3 pollen. These reports provide direct evidence that in petunia the biochemical mechanism for gametophytic self-incompatibility involves the ribonuclease activity of S proteins. Application of the methods used in petunia may elucidate the mechanism of S allele specificity in red clover.

5. Embryo development

Embryo ontogeny of red clover is classified as Onagrad type, variety *Trifolium* (Mackiewicz, 1965) . At 3 d after pollination intraspecific embryos are spherical but by 5 d are past the heart shaped stage and most are torpedo shaped (Armstrong, 1968). By 8 d cotyledons are becoming differentiated and at 11 d cotyledons are substantially enlarged. Embryos reach near full size by 14 to 17 d after pollination (Mackiewicz, 1965). An additional 7 to 10 d may be required to complete seed fill and dry down. As discussed in Chapter 2, interspecific hybrid embryos generally develop more slowly than intraspecific embryos. In crosses where hybrid embryos normally abort due to endosperm incompatibility, only globular heart shaped to torpedo shaped embryos are present at 12 to 15 d after pollination (Phillips et al., 1982; Sawai and Ueda, 1987).

Table 3.1

Gene names and recommended gene symbols for red clover.

Gene name	Recommended gene symbol
Disease resistance	
Bean yellow mosaic virus	N
Powdery mildew	Pm
Red clover mosaic virus	Rc
Rust	Ru
Dwarfism	
Dwarf plant	dw
Floral characters	
Purple-red flower	p
Rudimentary corolla	r
White flower	c
Leaf characters	
Elongated petiolule	ep
No leaf mark	m
Petiolulate leaflet	pl
Split leaflet	sp
Male sterility	
Male sterile 1	ms-1
Male sterile 2	ms-2
Miscellaneous	
Appressed stem hairs	hc
Appressed petiole hairs	hp
Glabrous stems	hg
Glabrous (basal internodes)	hgb
Hairless stipules	hs
Parallel spindles	ps
Round pollen	rp
Self fertility	Sf
Self incompatibility	S1, S2, . . .Sn
Synaptic mutant	sy
Nodulation	
Ineffective nodulation	i
Pigmentation	
Chlorophyll	Q
Albino	w
Yellow	y
Seed color	
Violet	v

Source: Quesenberry et al., 1991, with permission.

Fig. 3. 2. Leaf characters in red clover: (top to bottom): elongated petiolule, multiple leaflets, multiple normal leaflets with cornucopia leaflets, split leaflet.

6. Chromosomes, ploidy levels, and gene mapping

Although numerous simply inherited genes occur in red clover (e.g. Parrott and Smith, 1986a; Taylor and Snead, 1986) little genetic mapping has been accomplished. This is somewhat unusual for a diploid (2n = 2x = 14) species in which crossing procedures are relatively simple. It is perhaps due in part to a lack of a recognized marker collection, a situation which Quesenberry et al. (1991) attempted to rectify through a listing and description of recognized mutants. Twenty-nine gene symbols were recognized for red clover. Table 3.1 (reprinted from Quesenberry et al., 1991) lists these genes with their approved standardized gene symbol. A standardized nomenclature for genetic traits in red clover and a procedure for review of newly discovered genes was recommended. Following this voluntary procedure will insure that newly discovered genes are deposited in recognized germplasm banks. Several genes discovered in the 1920–40 era were not deposited in germplasm banks and are thus not currently available (Quesenberry et al., 1991). Fig. 3.2 shows four examples of leaf form variation in red clover. The gene for rudimentary corolla, r, is recessive to normal corolla (Taylor and Snead, 1986), and it has pleiotropic effects in the homozygous state. In addition to the shortened and crumpled corolla, the corolla shows little color and the homozygotes were completely male sterile and partially female sterile. It was expressed over a wide range of temperatures, but expression was slightly more pronounced at 30/27 C day/night than at 18/16 C day/night. This trait was not linked to leaf mark (M) or powdery mildew resistance (Pm).

Five additional reports have been published since the above summary. A mutation which suppresses flowering in red clover has been identified (Jones, 1991). Plants of this mutation were also more upright and densely tillering. The mutation was recessive to normal in the F_1, but the F_2 did not show the expected ratios. One back-cross of a F_1 plant with the mutant gave a ratio of flowering to non-flowering of 2:1 rather than 1:1. These results are consistent with the presence of a zygotic lethal factor originally present in the heterozygous form in the mutant non-flowering plant and tightly linked to the mutant gene. Segregating nonflowering plants always exhibited the characteristic mutant growth form. This gene was named development-influencing gibberellin (dig) and assigned the symbols F/f. Additional research has shown that this mutant will flower normally under long day when treated with gibberellin A3, A1, A7, or A5, but not when treated with A20, A8, or A13 (Jones and Thomas, 1993). Inheritance of the polymorphic isoenzyme leucine aminopeptidase in red clover has been reported (Maitre and Tarenghi, 1990). The enzyme was composed of five codominant alleles (LAP-284, LAP-296, LAP-2100, LAP-2104, and LAP-2112). All isozyme phenotypes were observed corresponding to the genotypes as expected. A single recessive gene which reduced petiole nonglandular trichomes from an average of about 20 per 1,000 epidermal cells to less than 1 per 1,000 cells has been reported

Table 3.2

Mean and range of cells in metaphase I configurations of eutriploid and aneutriploid plants from 4x × 2x crosses in red clover.

Chr. no.	No. cells	Chromosome associations							
		I		II		III		IV	
		mean	range	mean	range	mean	range	mean	range
21(5 pl.)	101	2.58	0–6	2.70	0–6	4.34	1–7	0	0
22(5 pl.)	101	2.32	0–7	2.92	0–7	4.26	1–7	0.27	0–1

Source: Taylor and Wiseman (1987).

Table 3.3

Relative lengths and arm ratios of the seven chromosomes of red clover.

Chromosomes	Mean Chromosome length %	Mean arm ratio (L/S)
I	8.35 ± 0.85	1.11 ± 0.26
II	8.57 ± 0.70	1.28 ± 0.19
III	7.28 ± 0.67	1.58 ± 0.34
IV	6.71 ± 0.46	1.17 ± 0.15
V	6.16 ± 0.43	1.19 ± 0.16
VI	5.86 ± 0.54	2.30 ± 0.55
VII	5.16 ± 0.50	2.25 ± 0.68

Source: Taylor and Chen (1988).

(Bowley and Lockie, 1989). This mutant did not affect frequency of glandular trichomes which averaged about 2.5 per 1,000 cells or stomata which averaged about 30 per 1,000 epidermal cells. The nonglandular trichomes are from 500 to 1,000 μm long and single celled whereas the glandular were about 100 μm but multicellular. The gene symbol proposed was *tn* and no data indicating linkage with any other gene was reported.

Germplasm releases of long stem (TP-LS) and multiple cotyledon (TP-MC) traits have recently been published (Taylor, 1995). The TP-LS germplasm has 112 mm long internodes compared to 68 mm for 'Kenstar', and is inherited in a quantitative manner (Bowley et al., 1984). The multiple cotyledon trait in the TP-MC germplasm is apparently controlled by recessive genes, but it has not been possible to maintain homozygous recessives even by inbreeding. Multiple cotyledon individuals may produce three or four cotyledons instead of the normal two.

Another hinderance to development of a genetic map is the lack of cytological stocks such as translocations or trisomics which might simplify isolation of a new gene to a specific chromosome. Production of a trisomic series in red clover was not thought possible due to an assumed strong triploid block (Anderson and Taylor, 1974; Parrott and Smith, 1984). However, when a 4x white flowered red clover plant was crossed with a group of 'Kenstar' plants using honey bees (*Aphis mellifera* L.) in a field cage, triploids were produced at a frequency of about 1 per 1000 flowers crossed (Taylor and Wiseman, 1987). Of 155 plants studied from this work, 119 were selfs of the white flowered female, 33 were triploid hybrids and 3 were tetraploid hybrids. Both eutriploid ($2n = 3x = 21$) and aneutriploid ($2n = 3x+1 = 22$) individuals were recovered at about equal frequencies. Although a frequency of 1 triploid per 1000 flowers crossed is rather low, the ability to use bees for pollination opened great possibilities for use of these triploids to produce trisomic lines.

Most frequent chromosome segregation at AI for either type was the most nearly balanced numbers (e.g. 11:10, 11:12, or 11:11), but configurations with only one additional chromosome (i.e. 13:8 or 14:8) were observed. The maximum of 7III was observed in 6.9% of the cells. Mean and range of MI pairing configurations of these triploids is shown in Table 3.2 (modified from Taylor and Wiseman, 1987). Of more interest is the segregation of chromosomes to the poles at AI. In the eutriploids 53.75% of the cells had 11:10 segregation, however 12.5% of cells had 13:8 segregation which would be expected to give rise to trisomic individuals if fertilized with a normal $n = 7$ gamete. This finding implied that recovery of trisomic individuals should be feasible from triploid × diploid crosses.

Chromosome numbers in 265 plants recovered from crosses of the above triploids × diploids included 2n = 14, 15, 16, 21, 26, and 28. About one third of the progeny were trisomic. To aid in identification of the seven different trisomics, determinations of the relative chromosome lengths and arm ratios were made for the seven chromosomes of red clover (Taylor and Chen, 1988). These results are reproduced in Table 3.3. Chromosome I had a satellite on the short arm. Chromosome IV and V were metacentric, III was submetacentric, and VI and VII were near telocentric (Fig. 3.3). These results were only in partial agreement with the karyotype of red clover reported by Angulo et al. (1972). The relatively small size of red clover chromosomes - all less than 2.9 microns (Kazimierski et al., 1972) likely contributes to some of the discrepancy between the reports of Angulo and that of Taylor and Chen. A $2n = 16$ karyotype was reported for a red clover accession from Mongolia (Zhang et al., 1993). This karyotype did not include an obvious satellited chromosome and was composed of seven more or less metacentric chromosomes plus one submetacentric chromosome.

Due to the heterozygous nature of red clover, Taylor and Chen (1988) were not able to associate definite phenotypes with individual trisomics. Stability and transmission of the trisomic condition has also not been reported. Double

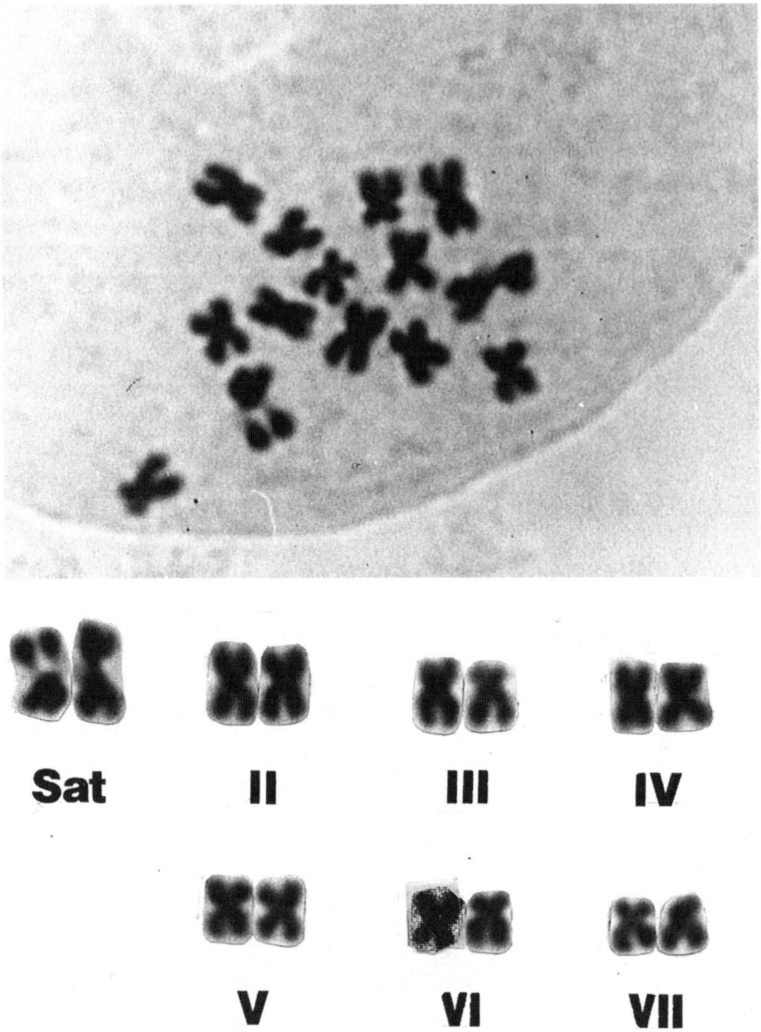

Fig. 3. 3. Karyotype of red clover (2n = 14).

trisomic plants (2n = 2x+2 = 16) were also found in this research. Use of the trisomic stocks for specific gene mapping would be simplified if specific phenotypes were associated with each trisomic. The seven trisomic red clover stocks have been released in a self-fertile background (Taylor, 1992).

Various aspects of methods for producing tetraploid red clover and the advantages of each are covered in Chapter 13. Differences in chromosome pairing among tetraploid lines derived by different methods are also discussed. In general, analyses of chromosome pairing at MI of microsporogenesis in

Fig. 3. 4. Octoploid red clover plant.

tetraploid red clover has shown between 45 and 75% of the chromosomes associated as bivalents (Povilaitis and Boyles, 1956; Schwer and Cleveland, 1972a; Taylor et al., 1976; Taylor and Giri, 1983; Taylor and Wiseman, 1987; Tofte and Smith, 1989). Both chain and ring quadravalent forms were observed (Schwer and Cleveland, 1972b). Nevertheless, the finding of a mean of at least 50% of the chromosomes of autotetraploid red clover paired as bivalents implies that even in newly produced autotetraploid red clover, sterility due to cytological aberrations is not likely to be a major factor. The octoploid level of red clover appeared to produce physiological imbalances resulting in large thickened leaves and petioles (Taylor et al., 1976) (Fig. 3.4). These octoploid plants were sterile or died before flowering. Pentaploids (2n = 5x = 35) also occur in red clover (Taylor and Wiseman, 1987; Strzyzewska, 1974) (Fig. 3.5). Taylor and Wiseman speculated that their pentaploid plant resulted from the union of a 2n (4x) female gamete with a normal (x) male gamete, since the seed from which this plant arose was harvested from a tetraploid female. This plant was morphologically similar to other tetraploids and had pollen stainability of 75%. The pentaploid plant reported by Strzyzewska only had 46% pollen stainability and had more rounded leaves. The production of fertile hexaploid red clover remains a possibility (Taylor and Wiseman, 1987).

7. Evolutionary trends

As discussed in Chapter 2, the fact that red clover is the only species in section *Trifolium* subsection *Trifolium* with x = 7, while all other species in this and other apparently closely related subsections (i.e. subsections *Intermedia* and subsection *Alpestria*) have x = 8, makes understanding of the chromo-

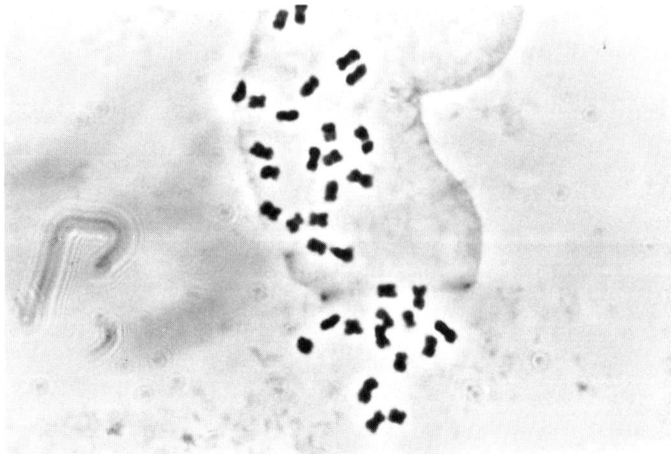

Fig. 3. 5. Somatic chromosomes of pentaploid red clover (2n = 35).

somal evolutionary pathways difficult. Except for one abstract of a Russian work (Abramova and Pyzhenkova, 1989) which reported 2n = 16 aneuploid individuals in the cultivar Sivoritskii, and the report of Zhang et al. (1993), no red clover types with karyotypes evolving between x = 8 and x = 7 are known. Abramova and Pyzhenkova (1989) state that the 16 chromosome form did not differ in general karyotype. Further independent confirmation of these reports are needed, but these may represent intermediate forms between perennial species and red clover.

Section *Trifolium* contains species of x = 5, 6, 7, and 8 (Zohary and Heller, 1984). In *Trifolium*, polyploidy is generally associated with perenniality, rhizomatous, or stoloniferous rooting habit, and self-incompatibility (Taylor et al., 1979). Except for *T. dubium* Sibth, (2n = 4x = 28) (Taylor et al., 1983), naturally occurring polyploids of x = 7 are not known in *Trifolium*. This species also represents the only known polyploid annual *Trifolium* spp., with the possible exception of *T. dichotomum* Hook & Arn. (2n = 32). Evidence presented in Chapter 2 also suggests that red clover is more closely related to annual (*T. diffusum*) than to perennial (*T. alpestre* or *T. medium*) species. Even so, red clover seems to differ from *T. diffusum* by multiple chromosome rearrangements. It may be that *T. diffusum* evolved by different types of chromosomal changes, maintaining x = 8, while red clover evolved to x = 7. The identification of the S_f allele in red clover makes plants carrying this gene similar in reproductive habit to *T. diffusum*. Thus, it would appear that red clover is a more evolved species than other perennial species in Section *Trifolium*. Additional analyses of chromosome pairing relationships in red clover × *T. medium* or red clover × *T. alpestre* hybrids may clarify some of these phylogenetic relationships. However, at present red clover appears to be more distantly related to the perennial species. A discussion of red clover gene pools is included in Chapter 2.

References

Abramova, L.I. and Z.S. Pyzhenkova. 1989. Polymorphism of chromosome number in red clover (*Trifolium pratense* L.) varieties from the collection. (In Russian) Sbornik Nauchnykh Trudov po Prikladnoi Botanike Genetike i Selektsii. 124:4–7.

Anderson, M.K. and N.L. Taylor. 1974. Effect of temperature on intra- and interspecific crosses of diploid and tetraploid red clover, *Trifolium pratense* L. *Theor. Appl. Genetics.* 44:73-76.

Anderson, M.K., N.L. Taylor and J.F. Duncan. 1974. Self-incompatibility genotype identification and stability as influenced by inbreeding in red clover (*Trifolium pratense* L.). *Euphytica.* 23:140–148.

Angulo, M.D., A.M. Sanchez De Rivera and F. Gozalez Bernaldez. 1972. Estudios chromosomics in el genero *Trifolium*, VI. *Lagascalia.* 2:3–13.

Armstrong, K.C. 1968. Cytogenetic and embryological studies of the interspecific hybrid *Trifolium pratense* L. × *T. pallidum* Waldst. and Kit. Ph.D. dissertation. The Pennsylvania State Univ., University Park, PA (Diss. Abstr. 29/05-B, 68–15114).

Bekuzarova, S.A., B.K. Mamsurov and M.M. Papina. 1987. Selective fertilization in red clover. Selektsiya i Semenovodstovo, USSR 2:14–15.

Bowley, S.R. and S.M. Lockie. 1989. Genetics of nonglandular stem trichomes in red clover (*Trifolium pratense* L.). *J. Hered.* 80:472–474.

Bowley, S.R., N.L. Taylor and P.L. Cornelius. 1984. Phenotypic recurrent selection for stem length in 'Kenstar' red clover. *Crop Sci.* 24:578–582.

Cleveland, R.W. 1985. Reproductive cycle and cytogenetics. In N.L. Taylor (ed.) *Clover Science and Technology. ASA-CSSA-SSSA Mono.* 25:71–110, Madison, WI.

Denward, T. 1963. The function of the incompatibility alleles in red clover (*Trifolium pratense* L.). III. Changes in the S-specificity. *Hereditas.* 49: 285–329.

Evans, A. M. 1962. Species hybridization in *Trifolium* II. investigating the pre-fertilization barriers to compatibility. *Euphytica.* 11:256–262.

Hermsen, J.G.T. 1984. Mechanisms and genetic implications of 2n-gamete formation. *Iowa State J. Res.* 58:421–434.

Heslop-Harrison, J. and Y. Heslop-Harrison. 1982. Pollen stigma interaction in the leguminosae: constituents of the stylar fluid and stigma secretion of *Trifolium pratense* L. *Ann. Bot.* 49:729–735.

Hindmarsh, G.J. 1964. Gametophyte development in *Trifolium pratense* L. *Aust. J. Bot.* 12:1–14.

Huang, S., H. Lee, B. Karunanandoa and T. Kao. 1994. Ribonuclease activity of *Petunia inflata* S proteins is essential for rejection of self pollen. *Plant Cell.* 6:1021–1028.

Jones, T.W.A. 1991. Inheritance of a mutation which suppresses flowering in red clover. *J. Exp. Bot.* 42:1589–1594.

Jones, T.W.A. and A.M. Thomas. 1993. Flowering and gibberellins in a mutant red clover (*Trifolium pratense* L.). *Plant Growth Regul.* 12:11–16.

Kazimierski, T., E.M. Kazimierska and C. Strzyewska. 1972. Species crossing in the genus *Trifolium* L. *Genet. Pol.* 13:11–31.

Kendall, W.A. and N.L. Taylor. 1965. Growth of red clover pollen. *Crop Sci.* 5:241–243.

Kendall, W.A. and N.L. Taylor. 1969. Effect of temperature on pseudo-self-compatibility in *Trifolium pratense* L. *Theor. Appl. Genet.* 39:123–126.

Lee, H., S. Huang and T. Kao. 1994. S proteins control rejection of incompatible pollen in *Petunia inflata*. *Nature.* 367:560–563.

Leffel, R.C. 1963. Pseudo-self-compatibility and segregation of gametophytic self-incompatibility alleles in red clover, *Trifolium pratense* L. *Crop Sci.* 3:377–380.

Mackiewicz, T. 1965. Low seed setting in tetraploid red clover (*Trifolium pratense* L.) in the light of cytoembryological analyses. *Genet. Pol.* 6:5–39.

Maitre, J.P. and E. Tarenghi. 1990. Leucine aminopeptidase in red clover, *Trifolium pratense* L. leaves inheritance of a polymorphic locus. (In French) *C. R. Acad. Sci. Ser III Sci. Vie.* 310:485–488.

Mendiburu, A.O. and S.J. Peloquin. 1976. Sexual polyploidization and depolyploidization: some terminology and definitions. *Theor. Appl. Genet.* 48:137–143.

Mok, D.W.S. and S.J. Peloquin. 1975. Three mechanisms of 2n pollen formation in diploid potatoes. *Canad. J. Genet. Cytol.* 17:217–225.

Mousset-DeClas, C., F. Colas and J.F. Trontin. 1992. Variation in 2n gamete production in red clover (*Trifolium pratense* L.): effect of temperature and genotype. In A.Mariani and S. Tavoletti (ed.), *Gametes with Somatic Chromosome Number in the Evolution and Breeding of Polyploid Polysomic Species: Achievements and Perspectives.* Proc. Workshop, Perugia, Italy, 9–10 April 1992. Consiglio Nazionale delle Ricerche.

Nettancourt, D. de. 1977. Incompatibility in angiosperms. Springer-Verlag. Heidelberg, Germany. 230 pp.

Pandey, K.K. 1955. Seed development in diploid, tetraploid and diploid-tetraploid crosses of *Trifolium pratense* L. *Indian J. Genet. & Pl. Brd.* 15:25–35.

Parrott, W.A. and R.R. Smith. 1984. Production of 2n pollen in red clover. *Crop Sci.* 24:469–472.

Parrott, W.A. and R.R. Smith. 1986a. Description and inheritance of new genes in red clover. *J. Hered.* 77:355–358.

Parrott, W.A. and R.R. Smith. 1986b. Recurrent selection for diploid pollen formation in red clover. *Crop. Sci.* 26:1132–1135.

Phillips, G.C., G.B. Collins and N.L. Taylor. 1982. Interspecific hybridization of red clover (*Trifolium pratense* L.) with *T. sarosiense* Hazsl. using *in vitro* embryo rescue. *Theor. Appl. Genet.* 62:17–24.

Povilaitis, B. and J.W. Boyles. 1956. A cytological study of autotetraploid red clover. *Amer. J. Bot.* 43:169–174.

Quesenberry, K.H. and N.L. Taylor. 1976. Interspecific hybridization in *Trifolium* L., Sect. *Trifolium* Zoh. I. Diploid hybrids among *T. alpestre* L., *T. rubens* L., *T. heldreichianum* Hausskn., and *T. noricum* Wulf. *Crop Sci.* 16:382–386.

Quesenberry, K.H., R.R. Smith, N.L. Taylor, D.D. Baltensperger and W.A. Parrott. 1991. Genetic nomenclature in clovers and special purpose legumes. I. Red and white clover. *Crop Sci.* 31:861–867.

Sawai, A. and S. Ueda. 1987. Embryo development of the hybrid of *Trifolium medium* L. × *4x Trifolium pratense* L. *J. Japan. Soc. Grassl. Sci.* 33:157–162.

Sawai, A., S. Ueda, M. Gau and K. Uchiyama. 1990. Interspecific hybrids of *Trifolium medium* L. × 4x *Trifolium pratense* L. obtained through embryo culture. *J. Japan. Soc. Grassl. Sci.* 33:157–162.

Schwer, J.F. and R.W. Cleveland. 1972a. Diploid interspecific hybrids of *Trifolium pratense* L., *T. diffusum* Ehrh., and some related species. *Crop Sci.* 12:321–324.

Schwer, J.F. and R.W. Cleveland. 1972b. Tetraploid and triploid interspecific hybrids of *Trifolium pratense* L., *T. diffusum* Ehrh., and some related species. *Crop Sci.* 12:419–422.

Snow, R. 1963. Alcoholic hydrochloric acid-carmine as a stain for chromosomes in squash preparations. *Stain Technol.* 38:9–13.

Strzyzewska, C. 1974. Sib-mating in *Trifolium pratense* L. II. Cytogenetics of euploids, aneuploids, and polyploids. *Genet. Pol.* 15:255–291.

Taylor, N.L. 1982. Stability of S-alleles in a red clover double cross hybrid. *Crop Sci.* 22:1222-1225.

Taylor, N.L. 1992. Registration of seven self-fertile trisomic red clover genetic stocks. *Crop Sci.* 32:1519.

Taylor, N.L. 1995. Registration of three genetic marker stocks for red clover. TR-RC, TP-LS, and TP-MC. *Crop Sci.* 35:1241.

Taylor, N.L., M.K. Anderson, K.H. Quesenberry, and Linda Watson. 1976. Doubling the chromosome number of *Trifolium* species using nitrous oxide. *Crop Sci.* 16:516–518.

Taylor, N.L. and K. Chen. 1988. Isolation of trisomics from crosses of diploid, triploid, and tetraploid red clover. *Crop Sci.* 28:209–213.

Taylor, N.L. and N. Giri. 1983. Frequency and stability of tetraploids from 2X-4X crosses in red clover. *Crop Sci.* 23:1191–1194.

Taylor, N.L. and L.D. Snead. 1986. Inheritance of rudimentary corolla in red clover. *J. Hered.* 77:381–382.

Taylor, N.L. and E.O. Wiseman. 1987. Triploids and tetraploids from 4X-2X crosses in red clover. *Crop. Sci.* 27:14–18.

Taylor, N.L., J.M. Gillett, and N. Giri. 1983. Morphological observations and chromosome numbers in *Trifolium* L. Section *Chronosemium* Ser. *Cytologia.* 48:671–677.

Taylor, N.L., R.F. Quarles and M.K. Anderson. 1979. Methods of overcoming interspecific barriers in *Trifolium. Euphytica.* 29:441–450.

Taylor, N.L., K.H. Quesenberry and M.K. Anderson. 1979. Genetic system relationships in *Trifolium. Econ. Botany.* 33:431–441.

Tofte, J.E. and R.R. Smith. 1989. Cytology of bilaterally derived tetraploid red clover. *Crop. Sci.* 29:37–41.

Williams, R.D. and W. Williams. 1947. Genetics of red clover (*Trifolium pratense* L.) compatibility III. The frequency of incompatibility S alleles in two non-pedigree populations of red clover. *J. Genet.* 48:69–79.

Zhang, Z.P., L.H. Wu and Y.F. Kang. 1993. Karyotype analysis of *Trifolium pratense* and *Trifolium repens.* (In Chinese) Grassland of China. 3:65–66.

Zohary, M. and D. Heller. 1984. *The Genus Trifolium.* The Israel Academy of Sciences and Humanities. 606 pp.

4. Morphology and Physiology

An extensive treatment of morphology and physiology was published by Bowley and Taylor (1984). The present treatment will not attempt to duplicate the earlier treatment, but will use it as a framework for presenting the literature published since about 1984.

1. Morphology and anatomy

1.1. *Aerial system*

1.1.1. *Stem and crown*

Stems originate from crowns that mostly grow at or slightly above the soil level, but cultivars vary in this regard. The complex of axillary buds enlarges as the plant grows to form the crown. Stems of most North American cultivars are densely pubescent with non-glandular trichomes, whereas European and Chilean populations are almost glabrous, having only short appressed trichomes.

The dense pubescence of North American cultivars was thought to be the result of natural selection for insect resistance, but also may be a mechanism for stress tolerance. Pubescence, mostly of the non-glandular trichome type impedes hay drying and creates dust (Collins et al., 1991). Although absence of non-glandular petiole trichomes has recently been shown to be controlled by a single recessive gene, the trait responds more quantitatively on stems (Bowley and Lackie, 1989). It may be advantageous to grow non-pubescent types if insect attacks are not a problem.

Glandular trichomes also occur on stems and most plant organs of red clover (Retallach and Willson, 1988). Glandular trichomes are multicellular as opposed to the single cell nature of non-glandular trichomes. Both procumbent and erect glands which arise from epidermal cells are capitate with thick-walled stalks. Both kinds of glandular trichomes may be involved in plant defense.

Stems, along with leaves and petioles, possess epicuticular wax that under field conditions increases with age (Mosley, 1983). Wax is considered to be of

Fig. 4. 1. Variation in leaflet arrangements in red clover.

possible significance as a source of antifoaming compounds in bloat inducing pastures. Cuticular wax also imposes a barrier to water loss at low water contents. Treatments which remove or reduce the wax may increase drying rates (Harris and Dhanoa, 1984).

Stem lengths at maturity may vary up to 80 cm. An experiment to select for longer stems increased length in the Kenstar cultivar from 53 to 73 cm in the first harvest (Bowley et al., 1984). Internode numbers were increased only slightly (8.0 to 8.5 per stem). The increase in stem length was associated primarily with longer internodes, which in turn were associated with either more cells per internode, or a greater proportion of cells being able to attain their maximum size.

1.1.2. *Leaves*

Red clover has alternate leaves. The first true leaf is unifoliolate and succeeding leaves usually are palmately trifoliate. However, some plants in the British cultivar S-123 produce mostly unifoliolate, or bifoliolate leaves, at least early in their life cycle. Variants exist that have unusual numbers of leaflets as well as unusually shaped leaflets and petiole and petiolule arrangements (Fig. 4.1). Leaflets may have a whitish or yellow mark which is controlled by a number of alleles (See Chapter 3, Reproductive Biology, Genetics, and Evolution). Other details of leaflets are presented by Bowley et al. (1984).

1.1.3. *Heads and flowers*

The inflorescence of red clover is a terminal capitulum or head consisting of up to 300 flowers (florets). Heads are subtended by bracts, but are without an involucre. The terminal inflorescence is initiated in the axil of the penultimate bract. After initiation of the final bract, the remnant apical dome is transferred to become the least mature part of the inflorescence dome (Retallack and Willison, 1990a). The oldest florets rest basally and proximal to the penultimate bract, and develop with zygomorphic symmetry. Double-headed inflorences are commonly observed in many red clover cultivars. According to Retallack and Willison (1990b), when the inflorence meristem is still small, it becomes divided by a median furrow into two inflorescence meristems. The meristem proximal to the penultimate bract initiated florets earlier than its counterpart. The inflorescence apices were progressively reduced by the initiation of florets as hemispherical protuberances. At maturity, no vegetative tissue could be seen between the adjacent inflorences. Further insight into head development was provided by selection for multiple-parted flower heads. Six generations of selection increased the number of head parts from 1.88 to 7.40. The maximum number of parts observed was 12, associated with stem faciation, and the presence of multiple growing points within each stem, back to the base. Apparently, head formation is at its optimum at a population equilibrium of about 1.75 heads per stem (Taylor et al., 1985).

The individual florets are zygomorphic typical of legumes and consist of a calyx with 5 lobes; a corolla with 5 petals; a standard, 2 wings, and 2 fused keel petals. The 10 stamens are fused to encircle the pistil. The intact stigma possesses a smooth receptive surface with an impermeable cuticle which must be ruptured before pollen grains can germinate on it. The cuticle is ruptured when the flower is tripped. This mechanism provides a guard against premature selfing (Heslop-Harrison and Heslop-Harrison, 1983). However, the self incompatibility mechanism of red clover also inhibits selfing. At maturity red clover pollen is binucleate with three germinal pores, whereas diploid pollen has more than three (polymerous). Diploid pollen is somewhat cubical, whereas haploid pollen is cylindrical in cross section.

1.1.4. *Fruit*

Red clover seeds are botanically classified as pyxidia. They are kidney shaped, about 1.5 to 2.2 x 1 mm in size (see Chapter 15, Seed Production for a description.) Seeds normally have two cotyledons but up to six cotyledons have been observed. The multicotyledonous condition is under genetic control (Taylor, unpublished).

1.2. *Root system*

Red clover is generally considered to be primarily a taprooted species. However, the root varies considerably depending upon a number of factors including growth type, soil moisture, soil density, plant spacing, and genetic factors. The more upright type plants generally are taprooted, whereas low growing plants may be more fibrous rooted, and possess adventitious roots originating from the upper part of the crown. Adventitious roots are thought to impart longevity to individual plants. With moderate water supply, a random root

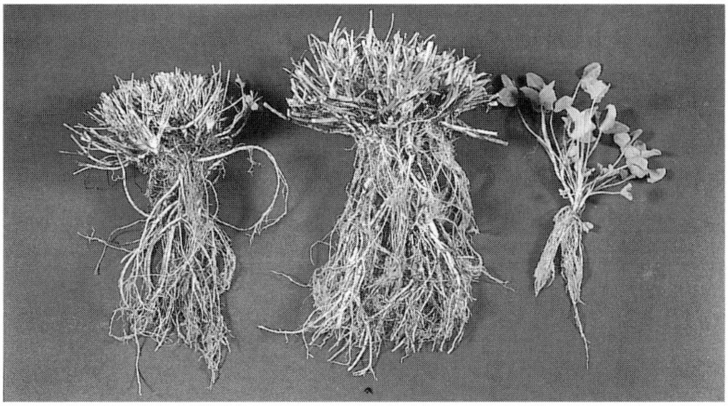

Fig. 4. 2. Branching in red clover roots.

system occurs, but excess water causes a departure from randomness (Fitter, 1986). Likewise, close spacing will result in a taproot, whereas plants in a widely spaced nursery usually will be much branched (Fig. 4.2). It is also possible to change the root system by selection (Smith, 1989). It is known that red clover stems will root at the nodes under optimum conditions, i.e. adequate moisture, and when procumbent stems are in close contact with the soil. Recently, a new cultivar, Astred, from Tasmania has been described that has persisted under pasture conditions for 16 years, retaining a ground cover of 55%. The clover is reported to achieve this persistence by virtue of its stoloniferous habit. The released cultivar is a descendent of CPI 48825, collected at Crato, Portugal in 1986 (Smith and Bishop, 1993).

2. Development phases and environmental effects

2.1. *Germination and seedling establishment*

Germination begins about three days after imbibition of water and emergence is epigeal. Imbibition may be limited by impermeable seed coats and seed dormancy particularly when seeds are freshly harvested. Hard seed coats may be overcome by various means of scarification, but care must be taken to avoid injury to the seeds. Seedlings of which the cotyledons have been broken off will usually not survive. Incidentally, ultrasonic seed treatment, presumedly first tried for scarification, resulted in increased yields. Thirty minute treatment increased yields over the untreated controls (5.17 vs 6.05t/ha) (Albu, 1986). Other methods of scarification have been described by Kendall and Stringer (1985). As the seed germinates, the radicle appears first, then the hypocotyl. Various dyes (including acetocarmin, alcian blue, congo red, and procion brown) may be used to distinguish the radicle from the hypocotyl.

All stains color the surface of the radicle but leave the hypocotyl unstained (Obrucheva, 1984).

Cotyledons of red clover expand in size during the first week after emergence and then enter senescence shortly after the development of the unifoliolate leaf. Larger seeds produce greater cotyledon area and the rate of expansion is closely related to seedling growth (Hur and Nelson, 1985). As seedlings grow and occupy more space, they thin themselves in accordance with the -3/2 thinning rule (Westoby, 1984). In crowded stands, seedlings compete with themselves and the smallest plants are the first to die.

3. Growth and reproduction

3.1. *Light interception*

Although canopy photosynthesis is reported to increase up to a lamina area index (LAI) of about 5.5 in the field, studies in a greenhouse (Joggi et al. 1983) suggest that net photosynthesis increases linearly with LAI up to 3.5 and declines at higher LAI, independent of cultivar and sowing density. Decrease in canopy photosynthesis above the optimum LAI is due to a higher proportion of old leaves with reduced photosynthetic capacity. After defoliation, about 2 weeks of regrowth are required to reach 95% light interception under optimum conditions. Maximum LAI declines in successive regrowth within each year but is otherwise not affected by age of stand. Lamina thickness is greater in thin than in thick stands (Maitre et al., 1985).

3.2. *Herbage and root growth*

Red clover accumulates herbage yield in a logarithmic fashion which becomes linear during stem elongation (Bowley et al. 1984). As with most forage legumes, total nonstructural carbohydrates (TNC) increase to a maximum at or just past full bloom. Root TNC levels decline during seed setting. TNC are used as an energy source for foliage regrowth. During regrowth the increase of root yield parallels the accumulation of herbage yield.

Herbage yield of red clover declines as the stand ages in accordance with its life cycle. Loss of stand is partly but not completely compensated by plant enlargement and increase in stems per plant. The stand eventually becomes too thin to be economically productive. At a very early age plant crowns begin to exhibit internal necrosis (browning, breakdown), which is more extensive in larger crowns. Apparently, no pathogen is involved in the early stages, but the necrotic region may be invaded eventually by various pathogens. Whether internal breakdown is the beginning of senescence is open to question (See Chapter 10 Persistence, for further discussion).

3.3. *Stem elongation and flowering*

Red clover requires long days to initiate stems and flowers. Stem elongation and flower initiation are thus affected by the same phenomena, day length. The two processes, however, can be separated by appropriate treatments. The critical day length to induce stem elongation may be shorter than that for flower induction. Longer days also produce longer internode lengths than shorter days, resulting in a delay in the time to flower. The combination of longer days to flower, longer stems, and more stems per plant may result in higher yields (Bowley et al., 1987; Puia et al., 1982). Increasing temperatures also may decrease the time for heading as it does in crimson and subclovers (Kendall and Stringer, 1985).

Red clover is generally considered to be a quantitative long day plant. Cultivars and plants within cultivars vary in this regard, depending upon origin. For example, in the Kenland cultivar, 10% of the plants did not flower when exposed to 14 hours day length but did flower under 16 hours (Bowley et al., 1987). Photoperiods of 9 hours may induce flowering in some cultivars (Reviewed by Kendall and Stringer, 1985). Red clover does not have a vernalization requirement, but cold treatments may accelerate flowering in some ecotypes from northern latitudes. Research conducted on diploid ecotypes from Norway showed that red clover had no vernalization or short-day requirement for flowering even at high latitudes (Lunnan, 1989).

The nature and ecological significance of the white or yellow marking (V-mark) on red clover (and other species) leaflets has long been the subject of speculation. The white marking region reflects 70% more visible radiation than the green portion of the leaflet, but is little different in reflectance of UV or IR. Light scattering apparently is due to air in the mesophyll cells, but the marking even when broad, did not appear to have a substantial impact in the energy budget of red clover leaflets (Feldhake, 1990).

4. Respiration

The respiration rates for red clover at stem elongation was 33 mg CO_2 plant/day (See review by Bowley and Taylor, 1984). The respiratory losses to gross photosynthesis ratio ranged from 0.6 to 0.8 in 2 year old red clover as compared to 0.35 and 0.45 in oats and potatoes respectively. Respiratory losses are greater in red clover because of use of assimilates for root respiration (Golovko, 1987).

5. Moisture

Differences in shoot-root weight and leaf area-root weight ratios suggested that alfalfa (*Medicago sativa*) was more drought tolerant than red clover despite lack of observable differences in leaf water potential or wilt susceptibility (Kendall and Hill, 1988). This study utilized the slant-board culture technique. The slant-board technique was also used by Kendall et al. (1994) to compare three cultivars of alfalfa and 3 cultivars of red clover at four levels each of water availability and temperature. Plants increased in dry weight and remained turgid in all treatments. Examination of shoot weights, root weights, shoot-root ratios, and number of roots showed significant differences among species except for shoot-root ratios. Cultivars within species were not significantly different. Shoot growth was inhibited more in red clover than alfalfa at low water levels and the reverse was true at high water levels.

6. Temperature

Data on temperature effects were obtained by Chugunova and Ronavova (1988) who compared early and late cultivars maintained at low (8°/6°C) and high (20°/18°C day/night) temperatures. The low temperature during early growth stages (up to unifoliate leaf appearance) followed by high temperature induced early maturity in the early cultivar. The early cultivar had greater fresh weight accumulation, higher rate of photosynthesis, and efficiency of its dark and light reactions and higher rate of dark respiration. No differences occurred when both cultivars were grown at the elevated temperature (20°/18°C day/night). In similar experiments, effect of temperature in early and late cultivars as expressed on ribulose-bisphosphate carboxylase was examined. Enzymatic activity was initially higher in the early cultivar, at 10°/4°C (day/night) but rose (compared with 20°/15°C treatment) to a greater extent in the late cultivar (Karpilova et al., 1988).

In slant-board studies the optimum temperature was 25 to 27 °C (based on root and shoot weights, respectively). The results are similar to those expected from field trials except optimum temperatures are higher than in other research (Kendall et al., 1994).

7. Mineral nutrition

The publication "Detecting Mineral Deficiencies in Tropical and Temperate Crops: Red Clover" (Peaslee and Taylor, 1989) covers in detail means of detecting deficiencies and nutrient responses.

7.1. *Phosphorus (P) and Potassium (K)*

Red clover is particularly intolerant of low P; not only are yields low but stands are difficult to obtain. Part of the difficulty is that red clover, in common with other forage legumes, is a poor competitor for P relative to forage grass species

Table 4.1

Effect of Brays extractable phosphorous and NH_40A_c extractable potassium on mean yields of red clover and timothy hay in Kentucky (Peaslee and Taylor, 1989).

Extractable P PPM	Relative yield (%)	Extractable K PPM	Relative Yield (%)
5	60	50	70
10	85	60	75
15	90	70	87
20	95	80	94
25	98	90	98

(Bagley and Taylor, 1987). In New Zealand, the critical P concentration for red clover in spring and early summer was determined to be 0.28%, which was intermediate between white clover and birdsfoot trefoil (Davis, 1991). The approximate P level (μ g p/ml) in soil solution at 90% of maximum yield for flatpea *(Lathyrus sylvestris*), was 0.074, for red clover was 0.104 and for birdsfoot trefoil was 0.141. The high internal P requirement and inability of red clover to achieve maximum yields at low adjusted P levels places it at a disadvantageous position relative to flatpea (Wright et al., 1984). Visual symptoms of P deficiency are primarily slow and stunted growth. Red clover can take up greater levels of P per unit of fresh root weight than associated perennial ryegrass, however, the ryegrass had higher fresh root weight (Mengel and Steffens, 1985). Thus, ryegrass is often more competitive for K than red clover, particularly where K level in the soil is low. Under these conditions ryegrass may be strongly competitive with red clover. Soil tests are available for both P and K. Parts per million of both elements and relative yields are presented in Table 4.1. Visual symptoms of deficiency include necrotic areas (flecked) near leaflet tips and mid-veins.

7.2. *pH*

Soil pH influences available Mn, Al, Mo, and P levels. Most growth and fewer disease problems occur when the soil pH is between 6.4 and 6.8 (Peaslee and Taylor, 1989) (Table 4.2). Red clover is very sensitive to Mn toxicity, but fortunately this nutrient decreases in availability when pH is below 5.7. An important exception to this may occur when greenhouse soils are steam treated under pressure.

Research has shown significant differences among and within red clover cultivars for growth in pH 4.5 solutions containing levels of Al ranging from 0

Table 4.2

Soil acidity and red clover production in 8 years of crop rotation (Peaslee and Taylor, 1989).

Soil pH	Relative yield[a]
4.7	12
5.0	21
5.7	53
6.8	98
7.5	100

[a] Expressed as $\frac{yield}{Maximum\ yield\ obtained}$ x 100.

Table 4.3

Relative yields of red clover plants grown in a greenhouse as influenced by soil pH and boron treatments (Highest yield = 100) (Peaslee and Taylor, 1989).

Soil pH	Relative yield of red clover tops Applied Boron, Mg/kg soil		
	0	0.5	2.0
5.3	0.54	0.68	0.60
5.8	0.74	0.86	0.78
6.3	0.62	0.93	0.98
6.8	0.52	0.98	0.94

to 100 μM (Baligar et al., 1987). A nutrient efficency ratio (ER) defined as mg of DW per mg of element in the shoot was developed to classify cultivars into efficient and inefficient utilizers of absorbed nutrients. The cultivars Kuhn and Proper I were recommended for moderately acid soils. Although low pH is known to reduce available soil P, research in Japan on low pH and low P soils showed that the effects of low soil pH and low P are independent of each other (Tanaka et al., 1984). When low pH and P conditions were combined, their adverse effects were increased. The adverse effects of low pH and low P declined with time elapsed after sowing.

Higher pH levels compensate for lower boron content (Table 4.3). Visual symptoms of boron deficiencies are reddish purple discolorations on margins of leaflets, gradually spreading to cover the entire leaf. These symptoms are often confused with potato leafhopper (*Empoasca fabae* Harris) damage, but usually may be distinguished by the latter having "V" shaped purple discolorations. (See Chapter 5 for additional discussion of pH effects on red clover production).

7.3. *Nitrogen*

Visual symptoms of deficiency include slow growth and pale green color. Rhizobium nodules are produced within the pH range of 4.5–6.0. (See Chapter 11, Rhizobium Relationships for further information on nitrogen fertilization aspects.)

8. Stress

Many agents, either absent or present, will induce stress in red clover (See Chapter 10, Persistence for other considerations of stress). Some environmental stresses are reviewed here.

8.1. *Salinity*

The safe tolerance level for growth of red clover is approximately 2.6 mmol NaCl/liter (See review by Bowley and Taylor, 1984). NaCl tolerant red clover selected on the basis of shoot growth in saline culture produced more dry matter than unselected control after 4 weeks of growth in sand culture with up to 175 mN NaCl. The tolerant line had a higher Ca^{2+} concentration than the unselected line (Ashraf et al., 1986).

8.2. *Ozone*

Heavily polluted air (though lower than that regularly occurring in the Los Angeles area), lowered net photosynthesis in all species tested including red clover. Growth and yield were also reduced (Reich and Amundson, 1985). Also, in Ontario, Canada, it was concluded from growth chamber research that ambient levels of ozone likely were injuring red clover foliage, reducing growth, and N_2 fixation (Ensing and Hofstra, 1982).

8.3. *Aluminum (Al)*

Differential cultivar and genotype response to aluminum was observed on a monmouth soil (26.2% Al saturation, pH 4.8) vs 2.8 Al saturation (pH 5.9)). Al stress reduced shoot and root growth in red clover cultivars (Campbell et al., 1990). In growth room studies, concentration above 50 μ m Al in nutrient cultures were sufficient to reduce shoot growth. Efficiency ratio (mg shoot dry matter per mg shoot element) for P declined with increasing Al. (Balegar *et al.*, 1987).

8.4. *Pesticides*

The possibility of pesticides (unspecified) being toxic and mutagenic was shown in Russia. Significant differences in pollen sterility of red clover occurred between plots with different intensities of pesticide use (Kurinnyi, 1983).

9. Modeling

Models to depict growth and development of red clover have been formulated. The model GROWIT has been shown to be valid for simulating growth of clovers both in pure stands and with companion crops (Smith and Loewer, 1985). A model adapted by Jacquard et al. (1987) to fit red clover consists of 2 sub-models. The environment is simulated, which determines growth and development and generates an average plant or a population of leaves and buds. Leaf senescence results from climate conditions and from the competition for light that limits LAI. At the same time, assimilates are migrated to a reserve component. However mineral nutrition and root growth are not considered (Jacquard et al., 1987).

10. Future research

Although morphology and physiology of red cover is now much better understood than in the past, much is still lacking. For example, under some conditions, stands that have flowered late in the first season and are not harvested (to remove apical dominance?) may be decimated by lower temperatures during the winter months. Apparently, lack of root reserves are involved in the loss of stands but further study is necessary because not all stands are depleted by such treatment.

Likewise, in the area of senescence much more information is needed. Senescence can be delayed in red clover by treatment with various hormones and cytokinens (Kendall and Stringer, 1985). The effect of these agents on internal breakdown and rotting of red clover roots, and their control of stand life is a promising area for investigation. (See Chapter 10 Persistence, for further considerations).

References

Albu, N. 1986. Increased yields in crop plants from seeds treated ultrasonically. Contributii Botanice, Universitatea Cluj Napoca, Gradina Botanica 1986:219–222.

Ashraf, M., T. McNeilly and A.D. Bradshaw. 1986. The response to NaC1 and ionic content of selected salt-tolerant and normal lines of three legume forage species in sand culture. *New Phyto.* 104:463–471.

Bagley, P.C. and N.L. Taylor. 1987. Evaluation of phosphorus efficiency in somaclones of red clover. *Iowa State J. Res.* 61:459–480.

Baligar, V.C., T.B. Kinraide, R.J. Wright, O.L. Bennett and M.D. Smedley. 1987. Aluminum effects on growth and P, Ca, and Mg uptake efficiency in red clover cultivars. *J. Plant Nutr.* 10:1131–1137.

Bowley, S.R. and S.M. Lackie. 1989. Genetics of nonglandular trichomes in red clover (*Trifolium pratense* L.). *J. Hered.* 80:472–474.

Bowley, S.R., N.L. Taylor and C.T. Dougherty. 1987. Photoperiodic response and heritability of the preflowering interval of two red clover (*Trifolium pratense*) populations. *Ann. Appl. Biol.* III:455–461.

Bowley, S.R., N.L. Taylor and C.T. Dougherty. 1984. Physiology and morphology of red clover. *Adv. Agron.* 37:317–347.

Bowley, S.R., N.L. Taylor and P.L. Cornelius. 1984. Phenotypic recurrent selection for stem length in 'Kenstar' red clover. *Crop Sci.* 24:578–582.

Campbell, T.A., N.J. Nuernberg and C.D. Foy. 1990. Differential responses of red clover germplasms to aluminum stress. *J. Plant Nutr.* 13:1464–1474.

Chugunova, N.G. and A.K. Romanova. 1988. Effect of temperature on growth, photosynthesis, and oxygenase activity of ribulose bisphosphate carboxylase/oxygenase in red clover. *Fiziologiya Rastenii.* 35:1071 1077.

Collins, M., E. Baker and N.L. Taylor. 1991. Trichome density effects on red clover hay drying. *Proc. Amer. For. Grassl. Council.* pp. 65–67.

Davis, M.R. 1991. The comparative phosphorus requirements of some temperate perennial legumes. *Plant and Soil.* 133:17–30.

Ensing, J and G. Hofstra. 1982. Impact of the air pollutant ozone on acetylene reduction and shoot growth of red clover. *Canad. J. Plant Path.* 4:237–242.

Feldhake, C.M. 1990. Spectral characteristics of the light band on red clover (*Trifolium pratense*). *Enviro. Exp. Bot.* 30:111–115.

Fitter, A.H. 1986. The topology and geometry of plant root systems: influence of watering rate on root system topology in *Trifolium pratense*. *Ann. Bot.* 58:91–101.

Golovko, T.K. 1987. Respiration and productivity of red clover, oats, and potatoes. *Fiziologiya I Biokhimiya Kul'turnykh Rastenii.* 19:334–342.

Harris, C.E. and M.S. Dhanoa. 1984. The influence of the epidermis on the drying rates of red clover leaf petioles and stems and of Italian ryegrass stems at low water contents. *Grass For. Sci.* 39:67–74.

Heslop-Harrison, J. and Y. Heslop-Harrison. 1983. Pollen stigma interaction in the Leguminosae; the organization of the stigma in *Trifolium pratense* cultivar Sabtoron. *Ann. Bot.* 51:571–584.

Hur, S.N. and C.J. Nelson. 1985. Cotyledon and leaf development associated with seedling vigor of six forage legumes. *Proc. Intern. Grassl. Congr.* 15:374–376. Kyoto, Japan.

Jacquard, P., B. Fleury and J.P. Maitre, 1987. Modelling a *Trifolium pratense* canopy in a pure stand: I. Foundations and description of the model. *Acta Oecologica Plantarum.* 8:211-236.

Joggi, D., V. Hofer and J. Nosberger. 1983. Leaf area index, canopy structure and photosynthesis of red clover (*Trifolium pratense* L.). *Plant, Cell, Envir.* 6:611–616.

Karpilova, I.F., A.K. Romanova, N.G. Ragimova and L.B. Pachepskaya. 1988. Effect of temperature on ribulose-bisphosphate carboxylase in the leaves of clover during early stages of development. *Fiziologiya Rastenii.* 35:41–50.

Kendall, W.A. and R.R. Hill, Jr. 1988. Some responses of alfalfa and red clover to water deficits in slant-board culture techniques. In Moutray, J.B. and J.H. Elgin, Jr. (eds), Report of 31st North American Alfalfa Imp. Conf. p. 59. Beltsville, MD.

Kendall, W.A. and W.C. Stringer. 1985. Physiological aspects of clover. In N.L. Taylor (ed.), Clover Science and Technology. 25:111–146. ASA-CSSA-SSSA Madison, WI. Mono.

Kendall, W.A., J.A. Shaffer and R.R. Hill, Jr. 1994. Effect of temperature and water variables on the juvenile growth of lucerne and red clover. *Grass For. Sci.* 49:249–269.

Kurennyi, A.I. 1983. The indication of environment pollution caused by mutagenic pesticides based on their gametocidal effect on plants. *Tsitol Genet.* 17:32–35.

Lunnan, T. 1989. Effects of photoperiod, temperature and vernalization on flowering and growth in high latitude populations of red clover. Norwegian *J. Agric. Sci.* 3:201–210.

Maitre, J.P., L. Assemat and P. Jacquard. 1985. Growth of red clover (*Trifolium pratense* L.) in association with Italian ryegrass (*Lolium multiflorum* Lam. subsp. italicum): II. Demography of organs, biomass and biological interference. *Agronomie.* 5:297–304.

Mengel, K. and D. Steffens. 1985. Potassium uptake of rye-grass (*Lolium perenne*) and red clover (*Trifolium pratense*) as related to root parameters. *Biol. Fert. Soils.* 1:53-58.

Moseley, G. 1983. Variation in the epicuticular wax content of white and red clover leaves. *Grass For. Sci.* 38:201–204.

Obrucheva, N.V., O.V. Antipova and A.C. Krovalev. 1984. Methods for distinguishing between radicle and hypocotyle in the embryo of germinating seeds of dicotyledons. *Bot. Zh.* (Leningr.) 69:1205–1209.

Peaslee, D.E. and N.L. Taylor. 1989. Red clover pp. 447–458. In Plucknett, D.L. and H.B. Sprague (eds). Detecting mineral nutrient deficiencies in tropical and temperate crops. Westview Press, Inc.

Puia, I., E. Pop, and M. Savatti. 1982. Photo-periodic response of some red clover ecotypes. Buletinul Institutului Agronomic. *Cluj Napoca, Agricultura.* 36:69–75.

Reich, P.B. and R.G. Amundson. 1985. Ambient levels of ozone reduce net photosynthesis in tree and crop species. *Science* 230:566–570.

Retallack, B. and J.H. Willison. 1990a. Floral development in Ottawa and Florex red clover, *Trifolium pratense* (Papilionoldeae:Legumlnosae). *Amer. J. Bot.* 77:1222–1230.

Retallack, B. and J.H. Willison. 1988. Morphology anatomy, and distribution of capitate glandular trichomes on selected *Trifolium* species. *Crop Sci.* 28:677–680.

Retallack, B. and J.H. Willison. 1990b. Common primordia and the double-headed inflorescence in 'Renova' red clover (*Trifolium pratense* L.), a papilionoid legume. *Acta Bot. Neerl.* 39:247–252.

Smith, R.S. and D.J. Bishop. 1993. Astred - A stoloniferous red clover. *Intern. Grassl. Congr. Proc.* 12: 421–423. Palmerston North, New Zealand.

Smith, E.M. and O.J. Loewer. 1985. Computer simulation of management and utilization systems. In N.L. Taylor (ed), *Clover Science and Technology.* ASA Monog. 25:355–364 ASA-CSSA-SSSA. Madison, WI.

Smith, R.R. 1989. Selection for root type in red clover. pp. 111–116. In Marten, G.C., A.G. Matches, R.F. Barnes, R.W. Broughan, R.J. Clements and G.W. Sheath (eds.), *Persistence of Forage Legumes.* ASA-CSSA-SSSA. Madison, WI.

Tanaka, A., K. Hitsuda and Y. Tsuchihashi. 1984. Tolerance to low pH and low available phosphorus of various field and forage crops. *Soil Sci. Plant Nutr.* 30:39–49.

Taylor, N.L., P.L. Cornelius and M.G. Long. 1985. Phenotypic recurrent selection for multiple-parted flower heads in red clover. *Crop. Sci.* 25:489–494.

Westoby, M. 1984. The self-thinning rule. *Adv. Ecol. Res.* 14:167–225.

Wright, R.J., M.C. Carter, T.B. Kinraide, and O.L. Bennett. 1984. Phosphorus requirements for the early growth of red clover, trefoil and flatpea. *Commun. Soil Sci. Plant Anal.* 15:49–64.

5. Management, utilization, quality and antiquality

1. Establishment

Red clover can be successfully established by a variety of different methods depending on intended use. As a group, the clovers are generally considered to have good seedling vigor and the larger seeded species like red clover usually emerge and compete well with companion crops or weeds. There are three primary methods of red clover establishment: 1) in cultivated soil as a pure stand, 2) in cultivated soil with a companion crop, and 3) as a renovation crop planted by various methods into an existing crop - usually a perennial grass pasture. Within each method various cultural practices may be employed.

1.1. *Pure stand establishment*

When red clover is established on cultivated soil as a pure stand, it may be planted either by broadcasting on the surface, with a drop seeder cultipacker, or with a shallow drill. Soil preparation is a key element to successful establishment. The seedbed should be level and firm with no large soil aggregates. Seed should be covered no more than 1 to 2 cm. Some have suggested that a small amount of N may be beneficial to red clover establishment. Recent unpublished research in Florida has shown little or no difference in total seasonal yield from addition of 50 kg/ha at establishment on a sandy soil (Quesenberry unpub.). There was a slight increase in first harvest yield due to the supplemental N. In more temperate areas of Europe, N applied in the establishment year generally has detrimental effects if a companion grass is included, but results are variable for pure stand establishment. If moisture conditions are deficient a shallow drilling tended to be the best sowing method, whereas broadcasting was best under wet conditions and on heavy soils (Oyen, 1991).

Herbicides are not generally used in the establishment of red clover, but the use of various graminicides in clover establishment has been investigated. Gallant 125 EE (haloxyfop) and Fusilade Super (fluazifop-butyl) were the most effective products studied (Bojas, 1988). In the Kirov region of Russia, use of the herbicides MCPB or 2,4-DB on seedling red clover at the 2-leaf stage decreased weed populations by 76–83% and increased yield of the companion barley crop (Kharkov and Starikov, 1985). Weed control increased light penetration to the legume and improved survival. The average establishment year dry matter yield of red clover with herbicide was 8640 kg/ha in pure stand and 7790 kg/ha under cover crops in alternate rows. Weed control can also be a major factor influencing establishment year yield and quality.

Season of the year may also have a major effect on red clover establishment. Depending on latitude, red clover may be planted in fall or spring. In the lower southeastern region of the U.S. a fall planting (September to October) is preferred. In areas from Kentucky north spring seedings are generally more successful. In these more northern latitudes, fall plantings may expose sensitive seedlings to crown rot (see Chapter 6 Fungal Diseases). Fall seedings also expose young plants to potential for heaving damage from freezing and thawing.

1.2. *Companion crop establishment*

Red clover is probably more frequently established with a companion crop than as a pure stand. If the ultimate intended crop is a pure stand, red clover may be planted with a annual cereal grain such as oat (*Avena sativa* L.), wheat (*Triticum aestivum* L.), barley (*Hordeum vulgare* L.) or rye (*Secale cereale* L.). Conversely, the companion crop may be a perennial pasture grass such as timothy (*Phleum pratense* L.) or tall fescue (*Festuca arundinacea* Schreb.). In temperate areas of the eastern U.S., the cereal grains are often established in fall and then red clover, with or without a perennial grass, is overseeded by broadcasting in late winter or early spring. The freezing and thawing of soil in this period creates good soil seed contact and germination occurs as soils warm. In Ireland, a seeding rate of 12 kg/ha of red clover with 2 or 4 kg/ha of timothy gave the most satisfactory stands (Frame et al., 1985). Mixtures of clover and grass increased total forage dry matter production compared to clover sown alone. Clover contribution to total yield increased at higher clover seeding rates. Similar seeding rates to those indicated above are recommended in most areas where red clover is grown. In Canada, inclusion of red clover as a companion crop with timothy resulted in lower timothy yields in the establishment year, but greater total yield. Even at the relatively low pH of 5.4 reported for this experiment, the mixture yielded more than pure stands of either species. Red clover in mixtures can improve overall forage quality (Lemieux et al., 1987).

In Poland, two diploid cultivars (a single cut and a double cut) and a double cut tetraploid cultivar were established with oats, spring barley for green fodder, spring barley for grain, or no companion crop. Establishment method had no major effect on growth and yield of the one cut variety, but establishment with a companion crop resulted in higher clover yields for the double cut type, especially where oat was the companion crop (Fordonski et al., 1988).

1.3. *Pasture renovation establishment*

Re-establishing red clover into a perennial grass sod is a recommended practice to improve overall pasture quality and yield. For such practice to be successful, the growth and water uptake of the established perennial grass must be suppressed to allow the emerging clover seedlings to compete for light and water. This suppression may be accomplished by disking the sod, close defoliation, herbicides, or combinations of these methods. Disking may loosen the soil surface causing rapid drying which may be detrimental for germinating seedlings. A rolling or packing of the soil at seeding can reduce this risk. If grazing is used to control grass growth, livestock must not be allowed to graze the grass so intensively that emerging clover seedlings are consumed. Damage from trampling may also reduce stands. A rest period after seedling emergence is required to allow the clover to become well established.

Grazing management when reestablishing clover into grasses can have dramatic effects on subsequent clover yields. In New Zealand, grazing perennial ryegrass pastures to 20 cm height at 3-week intervals over winter and early spring resulted in greater red clover survival and larger plant size than less intensive or less frequent grazing. Differences in red clover production among various grazing treatments and seeding rates were as much as 15 kg DM/ha/day (Campbell and Kunelius, 1984).

Herbicides used to control perennial grass when red clover is overseeded include dalapon, glyphosate, and paraquat. In one study in New York, dalapon applied a few days prior to planting did not improve red clover stands in comparison to unsprayed sod. Glyphosate applied at 2.2 kg/ha a.i. before seeding in spring or summer always increased red clover stands, but paraquat at 0.6 kg/ha was even more effective when applied in fall prior to a spring seeding. Improvement of the water content of the upper 20 cm of the soil profile was observed when the grass sod was treated with glyphosate (Trimmer and Linscott, 1986). In Florida, suppression of bahiagrass (*Paspalum notatum* F.) sod by paraquat or grazing improved percentage red clover from 18% in the untreated plots to 48 and 40% respectively (Kalmbacher and Martin, 1982).

Herbicides may be more effective in combination with seeding methods. In Japan, use of a slot seeder, disk-harrow, or no cultivation with various rates of paraquat were compared for mid- and late summer red clover establishment. Percentage ground cover of clover was highest with either cultivation method compared to no cultivation. Increasing rates of paraquat up to 1.0 l/ha resulted in further improvement of clover stand (Takeda et al., 1989).

Minimum tillage slot seeders have been used successfully in many countries for pasture renovation. The combination of slot seeders with herbicide suppression of the perennial grass has proven particularly effective for establishing red clover. Weed invasion was reduced and total weed free yields from application of a five inch band of paraquat over slot seeded rows were double those of the control without herbicide (Decker et al., 1986). Band width of the herbicide suppressant was important. Too narrow bands did not accomplish sufficient grass suppression while too wide bands resulted in excessive loss in grass DM yield (Haggar and Koch, 1983) In one study in the U.K., red clover slot seeded into a perennial ryegrass dominant sward without herbicide established well, whereas alfalfa did not. Slot seeding used in conjunction with herbicides can overcome several of the problems associated with renovation establishment of red clover. If conditions are favorable for legume seedling growth, slot seedings may be successful without herbicides, but some method of grass suppression will likely be needed (Guillard and Allison, 1987).

Season of the year may also be a critical factor for successful pasture renovation with red clover. In the middle latitudes of the eastern U.S. early spring is considered the ideal season, whereas in the lower southeast, fall seeding is recommended. Conversely, in Czechoslovakia early autumn was the most suitable season due to less competition from the perennial grass in fall than in spring (Jezikova and Michalec, 1988). In this study addition of red clover improved quality and increased Ca and Mg content of available forage.

2. Soil fertility

Optimum red clover growth is dependant on adequate levels of major and minor elements in the soil. Dry matter production is generally correlated to soil nutrient levels. Level of soil P is critical for legume establishment and inadequate P will often result in reduced stands. Available soil nutrient levels of 30 ppm P and 90 ppm K are generally considered as levels beyond which little or no increase in dry matter yields can be expected from additional increments of P and K fertilizer. On pastures which had received no P and K fertilizer for 11 years, rates of 45 kg/ha P and 120 kg/ha K increased clover yields by about 40% (Hrabe and Halve, 1989). Levels of P and K which need to be added to maintain productivity will depend on the type of utilization of the clover. If the clover is harvested for hay, all nutrients are removed and larger applications may be required to maintain soil fertility levels. Conversely, if the clover is used for grazing, most of the P and K and part of the fixed N in the plant is recycled to the soil to increase grass yields in mixed swards (Mengel and Steffens, 1985). Distribution of nutrients recycled in urine and feces may not be uniform and soil testing is still needed to determine fertility needs. Even when red clover-grass associations are harvested for hay, yields of clover-grass swards often equal pure grass swards fertilized with up to 200 kg/ha N (Shuppenies, 1988; Kiso and Kikuchi, 1988). In addition to recycling N from clovers to associated grasses, a plow-down effect of N released from decomposing red clover plants is frequently observed. Percentage N from clover recovered in a subsequent barley crop ranged from 56 to 85 on different soils (Muller, 1988). See Chapter 11, Rhizobium Relationships for additional discussion on N fertilization effects on red clover yield and N_2-fixation.

Soil pH influences the availability of Mn, Al, Mo, and P which subsequently can markedly effect red clover establishment and production. Optimum growth and fewer disease problems were reported to occur when pH was between 6.4 and 6.8 (Peaslee and Taylor, 1989). However, little response to raising pH above 5.5 was observed in Czechoslovakia (Habovstiak, 1989). Conversely, shoot growth of red clover in four U.S. Appalachian soils was positively correlated with soil pH over a range of 5.5 to 6.6 (Staley, 1993). These workers suggested that in the absence of effective naturalized rhizobia, improvement of rhizobial inocula could increase forage production by approximately 34% for some species on some of these toxic soils, even after the pH of the soils was increased to levels above 5.5. Red clover is very sensitive to Mn toxicity, but this symptom is usually only observed where greenhouse soils have been steam treated under pressure. See also Chapter 11 for additional discussion of effects of soil acidity on N_2- fixation and Chapter 4 for comments on effects of pH on nutrient deficiency symptoms.

Two of the major effects of low soil pH are increased availability of Al and decreased availability of P. Compared to 34 other plant species, red clover was more tolerant to Al than birdsfoot trefoil (*Lotus corniculatus* L.) and similar to white clover (*T. repens* L.) (McKenny et al., 1993). Other research has shown significant differences among and within red clover cultivars for growth in pH 4.5 solutions containing levels of Al ranging from 0 to 100 uM (Baligar et al., 1987). A nutrient efficiency ratio (ER) defined as mg of shoot DW per mg of element in the shoot was developed to classify cultivars into efficient and inefficient utilizers of absorbed nutrients. The cultivars Kuhn and Prosper I were recommended for moderately acidic soils. Although low pH is known to reduce available soil P, research in Japan on low pH and low P soils showed that the effects of low soil pH and low P were independent of each other (Tanaka et al., 1984). When low pH and P conditions were combined, their adverse effects were increased. The adverse effects of low pH and low P declined with time elapsed after sowing.

3. Management

3.1. *Persistence*

Harvest management of red clover affects both persistence over time and productivity. Regardless of whether red clover is spring or fall sown, adequate time must be allowed after the last defoliation in fall for ample accumulation of root carbohydrate storage reserves (Lang, 1985). Red clover needs about 45 days after a fall harvest to replenish carbohydrate stores before a hard freeze. In Canada, a two-cut system sometimes was superior to a three-cut system presumably because additional time was available for root reserve accumulation after the last harvest in the two-cut system (Coulman and Kielly, 1988). Similarly, increasing harvest frequency from a standard three cut system to four, five , or six cuts reduced DM and N yield and led to a more rapid reduction in plant number (Sheldrich et al., 1986). This reduction was true for both diploid and tetraploid cultivars. These workers concluded that even for silage use, strategies of more frequent cutting to improve digestibility were unattractive because of the yield and persistence reduction and relatively small improvement in digestibility. Some researchers have suggested that continuous grazing with a moderate stocking rate results in individual plants being defoliated once every two to three weeks. Thus based on the above findings regarding clipping frequency, only rotational grazing on a five week or longer schedule is recommended if red clover is grazed.

Research in Sweden has investigated the effect of nine fall cutting dates between 10 August and 5 October on overwintering of red clover (Anderson, 1986). This research showed that mid-September cutting resulted in poorest overwintering, presumably due to limited regrowth period before hard freeze. In other work from Sweden, a model to predict root TNC accumulation as a function of DM accumulation after fall cut has been developed. This model can be used to predict optimum time for last fall cut (Halling, 1988).

Sowing year harvest management may be the most critical for red clover stand maintenance. In Idaho, USA, spring sowing date (mid-April, early- and late-May) and harvest schedule (one or two cuts), and irrigation interacted to affect productivity and persistence in the second year (Hall and Eckert, 1992). With irrigation, the optimal management was mid-April sowing with first harvest at 80 days after emergence followed by a second harvest at early bloom. Without irrigation, late-May seeding with a single harvest was optimal. Seed heads of spring sown red clover must be removed by harvesting to eliminate apical dominance and allow development of strong rosettes before the onset of low temperatures. Failure to remove heads may result in loss of stands (Smith et al., 1985).

Cultivars may also interact with harvest management. Research from Czechoslovakia has shown that tetraploid and diploid cultivars may differ in optimal cutting time (Binderova, 1989). Harvest management in more northern latitudes may also be governed by cultivar selection (medium vs. mammoth type). The mammoth types are generally cut later and at later maturity stages than medium types. Even within red clover populations variability for flowering response is associated with persistence. In a large population of plants derived from crosses among 10 cultivars non-flowering plants persisted longer and had a greater proportion of roots arising from the crown (Sawai et al., 1986).

Genetic variability for cold hardiness (lowest temperature at which 50% of plants survive = LT50%) has been identified in red clover. The single cut (mammoth) type cultivar Norlac had the lowest (-11.6 C) LT50 followed by the single cut cultivar Altaswede with a LT50 of -10.4 °C (Choo and Suzuki, 1983a). However, cold hardiness of similar levels was found in four other double cut cultivars. The possibility of selection for increased cold hardiness, a trait which could improve winter persistence, is suggested. These researchers also found that plants must be at least 12 weeks old to be used in cold hardiness screening test (Choo and Suzuki, 1983b).

In summary, management for persistence of red clover stands must encompass consideration of time of seeding, irrigation or seasonal rainfall patterns, and cultivar (types), as well as allowance for adequate fall regrowth. All these factors, along with various diseases may interact to reduce stand persistence. Optimal management strategies will likely need to be developed for individual eco-geographic regions. See Chapter 10, Persistence, for additional discussion of this topic.

3.2. *Productivity*

A review of the recent literature reveals numerous examples of management of red clover for high hay yields, both as a pure stand and in mixtures with grasses. The frequency of reports from Eastern Europe and Russia suggests increased interest in these regions. In Poland, red clover in pure stand yielded 14,900 kg/ha of air dry forage in the first year and 10,500 kg/ha in the second year. Mixtures with meadow fescue and timothy yielded more than pure stands (Bawolski and Gawel, 1985). In Czechoslovakia, mixtures of red clover with timothy yielded 14,180 kg/ha DM when fertilized with 75 kg/ha N (Lesak and Sverakova, 1988). At seven locations in France, yields of red clover in mixtures with annual ryegrass or orchardgrass (*Dactylis glomerata* L.) fertilized with 150 kg/ha N were similar to pure swards of the grasses fertilized with 300 kg/ha N (Guy, 1989).

In the Southeastern U.S. research has shown that red clover interseeded into tall fescue with cool season annual clovers can increase total system production (Pederson and Brink, 1991). This research compared inclusion of red clover for increased summer forage production in Mississippi. Presence of red clover generally increased total yields by 1 to 2 Mg/ha, but addition of an annual clover to the system generally did not increase total yield. Research in Kansas, USA has shown that red clover planted in mixture with four different perennial grasses generally yielded equal to or up to 1 Mg/ha more than pure grass fertilized with 90 kg N/ha. Crude protein of the clover grass forage mixture was from 1 to 3% higher than pure grass with 90 kg N/ha (Barnett and Posler, 1983).

3.3. *Grazing management*

3.3.1. *Beef*

Incorporation of red clover into grass pastures can significantly increase beef animal production and performance. Adding red clover to pastures dominated by tall fescue in the cool season and by midland bermudagrass [*Cynodon dactylon* (L.) Pers.] in the warm season resulted in beef production of 597 kg/ha which was exceeded only by the same grasses plus N fertilizer (Mitchell et al., 1986). Development of a complete system to provide a year-around forage based diet for a beef cow-calf operation was studied in a 6-year experiment in Virginia, USA. In the spring, summer, and early fall, cows were grazed on one of two treatments (1) tall fescue-white clover, or (2) Kentucky bluegrass-white clover. Each of these forage mixtures was combined in a factorial arrangement with two paddocks of either (1) fescue-red clover, (2) orchardgrass-red clover, or (3) orchardgrass-alfalfa, which were used for hay, creep grazing by calves, and stockpiling for grazing by cows in late fall and winter (Allen et al., 1992b). Stockpiled fescue-red clover provided more ($P<0.05$) grazing days and required less ($P < 0.05$) hay fed to cows than stockpiled orchardgrass plus either red clover or alfalfa. The system of fescue-white clover combined with fescue-red clover required minimum inputs of harvested feed and maintained excellent stands during the six-year study (Allen et al., 1992b).

These same workers compared six systems of year-around forage for beef stocker calves in a five-year experiment. The systems were (1) tall fescue, (2) tall fescue-red clover, (3) tall-fescue-alfalfa, (4) barn-fed tall fescue hay, (5) barn-fed orchard grass-alfalfa, and (6) barn-fed tall fescue silage (Allen et al., 1992a). The feeding period was from late October to early April. Calves grazing stockpiled fescue plus red clover with no fertilizer had similar average daily gains as those grazing pure stockpiled fescue with 90 kg/ha N fertilizer applied in early spring and in early August (Allen et al., 1992a).

In New Zealand, research has shown that grazing management strategies exist which can increase both white clover and red clover content of perennial grass swards (Ray and Hunt, 1989). This work also showed that genotype differences can confer considerable additional advantages to the legume. The legume's superior light interception and N_2 - fixation capability allowed them to compete successfully with grasses. Other work in New Zealand has shown that addition of red clover to white clover-perennial ryegrass pastures to provide additional late summer and autumn production increased beef production by up to 13% annually (Brougham and Cosgrove, 1985).

In Europe, the advantages of incorporation of red clover into animal diets were also demonstrated. In Spain, addition of red clover and white clover to perennial grasses resulted in an increase of 25% in stocking rate (Sineiro et al., 1984). The primary advantage of red clover incorporation into grasses is a resulting increase in animal intake due to high protein, minerals and digestibility of the red clover with a resulting high feed conversion. The specific quality attributes of red clover will be influenced by stage of maturity.

3.3.2. *Dairy*

This relatively good forage quality makes red clover a forage with potential for dairy diets as well. In New Zealand, cows grazing Grasslands Hauma red clover gave more milk, protein, and fat in summer and autumn than cows grazing mixed ryegrass-white clover pastures (Brookes and Wilson, 1983). In Chile, pastures of red clover resulted in greater milk yield per cow and per hectare than those grazing birdsfoot trefoil, or alfalfa (Pedraza et al., 1988). A limitation to use of red clover for dairy rations is that optimum forage quality is obtained by harvesting at relatively immature growth stages which may adversely affect red clover persistence.

3.3.3. *Sheep*

Red clover can also be utilized effectively for sheep grazing. In New Zealand, three early flowering red clover cultivars showed interactions for yield under sheep grazing when combined with prairie grass (*Bromus willdenovii*) or perennial ryegrass. Grasslands Hauma was superior to Grasslands Colenso when associated with either grass. Red clover yields were greater under infrequent grazing (Hickey and Harris, 1989). In Ireland, sheep grazing a red clover-perennial ryegrass sward were shown to select for red clover. Selection was most pronounced in weeks 3 and 4 of a six week grazing cycle. Leaf laminae of red clover were selected in preference to other plant parts and were grazed throughout the canopy in contrast to the grass component (Laidlaw, 1983).

In summary, red clover is a desirable forage to increase beef yield and calf conception rate when added to grass pastures. It can serve as a component of dairy rations and can be used in sheep diets as well. The primary limitation to use of red clover appears to be due to limited persistence. Nevertheless, it remains the legume of choice for improving grazing animal performance on many grass based pasture systems throughout the world.

4. Forage quality

The "quality" of a forage can be estimated in a number of different ways all of which may (or may not) have relevance to the system in which the forage will be utilized. The ultimate measure of the quality of a forage is the performance of the animals consuming this forage; however, it is generally not practical to conduct *in vivo* evaluation of forage quality where there are large number of treatment variables. Thus, different laboratory methods have been developed to estimate various quality parameters. In much of the discussion which follows *in vitro* methods have been used to monitor the effects of treatments on changes in quality parameters. A section relating animal gains to quality is included to correlate *in vitro* and *in vivo* responses.

4.1. *Management effects on quality*

The three primary "management" variables that can effect red clover quality are growth stage (maturity), cutting height, and environmental stress. The two measures of forage quality most often evaluated are crude protein (CP) and *in vitro* digestible dry matter (IVDDM). Other laboratory values which may impact IVDDM are estimates of cell wall and cell content fractions.

It is generally accepted that IVDDM and CP of whole plants decrease with increasing maturity. This is the result of increasing percentage of cell walls (lignification) and a decreasing leaf to stem ratio. Leaves are generally more digestible than stems. However, in research by Buxton et al. (1985) red clover appeared to not decline in quality with advancing maturity as rapidly as other forages. Red clover declined in IVDDM at a rate of 8 g/kg/node (from top to bottom of a stem) vs 20 g/kg/node for alfalfa or birdsfoot trefoil. In addition, red clover stems had greater IVDDM than leaves until the time of flowering. Red clover was generally intermediate between white clover (highest) and alfalfa in quality. IVDDM of red clover declined more rapidly with maturity in stems than in leaves. However, CP decline with maturity was similar in leaves and stems. This research also illustrates the change in leaf-stem ratio with maturity. When stems were divided into upper and basal segments, leaf mass decreased and stem mass increased as plants matured. Nevertheless, red clover had the highest leaf-to-stem mass ratio (0.23) compared to 0.16 for birdsfoot trefoil, and 0.06 for alfalfa (Buxton et al., 1985). Red clover stems generally had lower cell wall content than alfalfa or birdsfoot trefoil at comparable maturity stages. Cell wall lignin was also lowest in red clover but there was little difference in cell wall lignin content between red clover leaves and stems (Buxton and Hornstein, 1986). Other research has confirmed a relatively high IVDDM for upper stem halves of red clover at the flower bud stage (Wilman and Altimimi, 1984). This research also showed a decline in IVDDM in the basal stem fractions. In this study digestibility of plant parts was strongly correlated with the proportion of cell contents, which in turn was positively correlated with N% (CP). Digestibility of the cell wall was negatively correlated with proportion of lignin in cell walls.

Work in Sweden has shown that one reason for the decline in IVDDM with maturity is that the degradibility (digestibility) of non-starch polysaccharides (NSP) decreased from 76 to 51% with maturity (Aman and Nordkvist, 1983). Regardless of maturity stage, lignin was not degraded, starch was degraded 100%, and CP about 80%. Thus the decrease in IVDDM appears to be due to a combination of increased lignin and a decrease in degradability of NSP. The greatest decrease in *in vitro* degradability with maturity was for the polysaccharide xylose. Further research using plants fractionated into leaves, petioles, and stems showed that the degradibility of NSP was highest in leaves and lowest in stems (Aman, 1985). Thus an increasing proportion of stem also results in a increasing proportion of NSP which is poorly degraded.

Since several of the above cited studies have shown that stem bases are less digestible and lower in CP than upper stem parts, it follows that cutting height will impact total forage quality. However, higher cutting heights will reduce overall DM yield. In Chile, IVDDM increased with cutting height, but digestible energy and CP production per hectare increased as cutting height decreased due to increased DM yield at the lower cutting height (Silva et al., 1988). Caution must be noted with such calculations (multiplication of DM yields × IVDDM) since no comparisons of animal performance on the forages of different IVDDM where included.

Environmental stress and seasonal changes may alter forage quality parameters. Although drought will likely result in marked reduction in DM yield, quality of forage produced under drought is usually higher. Acid detergent fiber (ADF) of drought stressed red clover was reduced by 30 to 48% compared a well watered control (Peterson et al., 1992). Overall improved quality in the drought-affected red clover was related to greater leaf: stem ratio, delayed maturity, and higher leaf and stem quality parameters. Herbicides and growth regulators may also impact red clover quality. Application of chlormequat chloride and daminozide decreased cell wall components and increased CP, but mefluide and cytokinin increased cell wall components (Buck et al., 1989).

Red clover grown in mixtures with perennial grasses such as tall fescue may be allowed to accumulate in early fall for late fall and winter use (forage stockpiling). Under this type management N (CP) content, declined after an initial frost but then stabilized after about 30 days (Collins and Taylor, 1984). However, IVDDM showed a continual decline from a maximum of near 70% and reached a minimum of about 51%. Substantial total losses from weathering may occur with this type management since yield declines more after frosting than does quality (Collins, 1983).

Increase in lignin content with maturity is the major factor contributing to decreased IVDDM in most forages. There is often a strong correlation between lignin concentration and cell wall digestibility, suggesting that lignin complexes inhibit digestion of other cell wall constituents (Buxton and Russell, 1988). In red clover, acid detergent lignin increased from 6.2 to 11.2% between 18 June and 15 July as stem length increased from 15 to 82 cm. (Kuhbauch and Bestajovsky, 1989). The lignification of red clover vascular bundles is stronger than in alfalfa, but remains confined to the vascular bundle, whereas alfalfa lignification spreads from the vascular bundles into surrounding parenchyma tissue and penetrates increasingly larger proportions of the stem.

Similar to other legumes and grasses the phenolic components of red clover cell wall lignin were p-coumaric and ferulic acid, although red clover had 2

Fig. 5. 1. Well cured round bales of red clover-grass hay in Kentucky.

to 10 times lower concentrations of these compounds than various grasses (Eraso and Hartley, 1990). Other work has shown that alkaline-nitrobenzene oxidation of red clover lignin yielded p-hydroxy benzaldehyde, vanillin, and syringceldelhyde as the major phenolic aldehydes (Kondo et al., 1986).

4.2. *Haying or ensiling effects on quality*

Drying of red clover for preservation as hay can result in loss of total dry matter, minerals, amino acids, and IVDDM. Research from Japan has shown that free proline and total proline of red clover sun-cured for five days increased by 37 and 1.9 fold compared to levels in fresh forage (Igarashi et al. 1985). Content of total methionine, total lysine, and total tyrosine decreased by 23, 24, and 20% under the same drying conditions. Forced air drying at 20 °C for 0 to 126 h also affected amino acid content. Free proline content was 0.10, 1.94, 4.61, and 8.44 mol/g DM after 0, 27, 44, and 126 h of drying (Igarashi and Yasui, 1989).

Wetting by rainfall can also cause deterioration in red clover hay quality. Rainfall during hay curing increased NDF, ADF, cellulose, and ADL concentration with a resultant decrease in IVDDM (Collins, 1985a). Concentrations of several minerals essential for ruminants in red clover hay are reduced by rainfall during field curing (Collins, 1985b).

Various methods may be used to speed up the drying process to help avoid rainfall and thus reduce DM and quality losses. The major types of methods for increased drying are mechanical, chemical, or combinations of both. In the Canadian humid Maritime region red clover sprayed with 5 kg/ha K_2CO_3 at cutting dried faster than untreated forage (Narasimhalu and Choo, 1983). Hay treated with a combination of mechanical conditioning and K_2CO_3 dried faster than either treatment alone (Clark et al., 1989). Heavy windrows and poor drying conditions (high relative humidity) lessened the value of either treatment. Thus, neither treatment can compensate for cool, cloudy weather or heavy dense windrows. In Kentucky (USA) similar results were reported. Mechanical conditioning and treatment with K_2CO_3 increased DM concentration from 65.3 to 78.6 g/kg at 72h after cutting compared to untreated forage (Collins, 1989). Fig. 5.1 shows well cured round bales of red clover hay in Kentucky. In Japan crimping rollers resulted in faster drying rate for red clover than crushing rollers (Kawakami, 1982). Emulsified methyl esters of long chain fatty acids plus K_2CO_3 have also been used to speed drying of red clover (Johnson et al., 1984).

In one study from Korea, laboratory dried hay treated with K_2CO_3 or KOH tended to have lower fiber content than the control. Extent of NDF digestion *in vitro* was increased at 12, 24, and 48h of incubation. Both chemical treatments decreased the proportion of indigestible NDF (Hong et al., 1989). Thus it appears these chemical treatments may increase red clover hay quality.

An alternative approach to increase drying rate of red clover is to genetically reduce trichome density on stem internodes (Baker et al., 1993). In this research, pubescence density and conditioning treatments both affected overall mean drying rate. Sparsely pubescent shoots lost moisture 56% more rapidly than densely pubescent shoots. Conditioned shoots dried an average of 54% more rapidly than unconditioned shoots. Smooth, conditioned shoots dried at more than twice the rate of highly pubescent, unconditioned shoots. Levels of nonglandular stem trichomes in red clover is a trait that can be manipulated genetically (Bowley and Lackie, 1989).

Conservation of red clover as silage may also impact quality. Red clover will usually not ensile well as fresh forage due to low DM content. Consequently, field wilting or treatment with formic acid may be used to improve ensiling. The rate of formic acid necessary is 0.3 to 0.5% of ensiled material (Tagaev et al., 1985). Either benzoic acid or a mixture of formic and benzoic acid may be used. In Czechoslovakia, these additives increased red clover silage DM, reduced ammonia-N, and eliminated butyric acid (Chlevickas et al., 1990). In Chile, silage DM and CP content was higher when treated with sodium formate than in the control (Siebald et al., 1989). *In vitro* digestibility of treated silage was higher than untreated at first cut but not at second. The treatment reduced pH and increased silage lactic acid content and liveweight gains of steers.

Research in Australia, England, Poland, and France has shown that red clover silage can be used effectively for dairy and sheep production. In one study, cows in weeks 3 to 20 of lactation were fed 7 kg daily of barley, ground maize, and soybean and ad libitum access to red clover silage or, red clover silage weeks in 3 to 10 followed by perennial ryegrass silage weeks in 12 to 20, or grass silage for the entire period (Thomas et al., 1985). Cows fed red clover only consumed 1.22 times the DM and 1.09 times the digestible energy as those given perennial ryegrass. Cows given red clover silage yielded

Fig. 5. 2. Beef stocker steers grazing red clover grass pasture.

1.9 kg/day more milk than those on perennial ryegrass. However, this milk had lower fat concentration (37.2 vs 41.2 g/kg) than milk from cows on perennial ryegrass. In another study in Australia, research showed that grazing clover pastures between morning and afternoon milking supplemented with a combination of maize and red clover silage sustained high milk yields (Moran and Wamungai, 1992). In France, cows consuming 60% red clover silage and 40% maize silage, produced as much milk, which had lower fat content than those consuming maize silage alone (Hoden et al., 1987). All these studies show that red clover silage can be used effectively as a component of dairy diets, but that special care must be taken in preservation.

4.3. *Animal performance as a measure of quality*

Since the ultimate measure of forage quality is animal production, the studies cited below are examples of weight gain which may be expected from red clover. Results of most of these studies show that animal performance is primarily determined by daily feed consumption (intake), which is a function of rate of passage and disappearance of feed through the digestive tract. A primary determinant of rate of passage is speed of digestion which is closely related to fiber content of the feed (lignification). Thus, as previously discussed, maturity of red clover forage will impact overall animal performance. Fig. 5.2 illustrates beef stocker steers grazing high quality red clover - grass pastures.

Rates of NDF and ADF disappearance of red clover were greater than orchardgrass when compared by *in vitro* nylon bag digestion (Bowman and Firkins, 1993). Carboxymethyl cellulose activity of red clover was higher indicating that red clover supported a greater total particle associated bacteria (PAB) population in the rumen than did orchardgrass. A difference in colonization rate of forage particles by cellulolytic PAB relative to total PAB was observed which may account for the increased rate of digestion for red clover. Research in Czechoslovakia has also shown that intake of red clover hay was higher than intake of grass hay (Mika and Paul, 1988). Voluntary intake in this study was positively correlated with DM digestibility and CP.

Similar results for feeding of red clover hay were obtained for other classes of livestock. Though it is not usually fed to non-ruminants, a study from China reports substitution of 0, 5, 10, and 15% red clover meal to replace wheat bran and rapeseed meal in pig diets (Liu et al., 1991). The four diets did not differ significantly in daily gain and feed:meat ratio. In Korea, DM intake of goats increased as the proportion of red clover in a clover-orchardgrass mixture increased (Lee et al., 1987). Digestible dry matter intake was positively correlated to CP content and negatively correlated to NDF.

Red clover appears to be a particularly good forage and is highly preferred by red deer (Hunt and Hay, 1990). Red deer fawns reared on hinds grazing pure red clover swards during lactation had significantly higher growth rates and weaning weights compared to controls grazing perennial ryegrass/white clover (Niezen et al., 1991). Another experiment was conducted to evaluate calf and hind performance on red clover vs. perennial ryegrass-clover in the period of 1 to 3.5 months post-parturition (Nizen et al., 1993). In this study, hinds on red clover had higher voluntary intake and calves had higher weight gain than those on the ryegrass-clover mixture (430 vs 330 g/day). Further work evaluated growth of red deer weaners on red clover vs perennial ryegrass-white clover (Semiadi et al., 1992). In this study, the test stags and hinds were rotationally grazed on the two separate treatment in fall and spring, but were all grazed on perennial ryegrass with supplemental hay in winter when red clover was dormant. Animals grazing red clover had higher voluntary intake in the autumn and spring than those on ryegrass. At the end of the experiment (when animals were 12 months of age) stags on red clover were 7 kg heavier than those on ryegrass and hinds were 3 kg heavier. All stags on red clover attained minimum slaughter weight (92 kg) whereas only 75% of those on ryegrass reached this weight. Slaughtered carcasses from red clover stags were heavier and had higher dressing percentage with no difference in fat deposition. Additional research with red deer confined indoors in metabolism pens showed higher voluntary intake for red clover than for perennial ryegrass (Freudenberger et al., 1994).

5. Antiquality factors

The primary antiquality factors associated with red clover are: (1) incidence of bloat caused by excessive rumen foam trapping gases from fermentation, (2) adverse reproductive behavior attributed to isoflavone compounds, and (3) excess salivation caused by consumption of hay contaminated with certain fungi. Problems with these factors can be avoided by proper crop and animal management at critical times. However, failure to give attention to these factors can lead to loss of livestock.

6. Bloat

Bloat problems from grazing red clover usually occur when the pasture is young, lush, and rapidly growing, and/or when hungry animals are moved suddenly to a sward containing a high percentage of red clover. Pastures containing greater than 50% grass on a dry weight basis will usually not cause bloat. A general review of recent research related to bloat problems with forage legumes was presented by Bush and Burton (1994). It has been shown that pasture legumes such as birdsfoot trefoil which contain tannins do not cause bloat. *In vitro* tests have shown that tannins bind proteins and prevent foaming (Kendall, 1966). Although 11 species of *Trifolium* have been shown to contain varying levels of tannins (Fay and Dale, 1984) none of these will hybridize with red clover. Genetic transfer of tannins to red clover will likely only be possible with improved understanding of tannin biosynthesis and improved genetic transformation techniques.

Chemical treatments have also been proposed as a remedy for bloat. In a pilot controlled feeding study, monensin controlled-release compounds (CRC) reduced the incidence of bloat in cows selected for high susceptibility to bloat and grazing red clover, while milk and protein yield increased (Lowe et al., 1991). Following the pilot study, six large scale field experiments with a total of 368 lactating dairy cows were conducted in Australia and New Zealand. When the data were pooled over the four experiments in which bloat occurred, it was concluded that bloat was reduced by the monensin treatments. Daily milk yield increased in all experiments from a mean of 17.7 in the untreated groups to 18.8 kg/head/day in the monensin CRC treated group.

7. Isoflavones

Diets high in red clover content have been associated with infertility in sheep. This infertility arises from a variety of reproductive malfunctions apparently due to ingestion of the isoflavone formononetin, a compound which occurs naturally in red clover. Formononetin is metabolized in the rumen to the active oestrogenic compound equol which is subsequently absorbed into the bloodstream (Moseley et al., 1984). The principal reproductive effects appear to be reduced conception due to a reduced oestrus incidence (Kelly and Shackell 1982; Kelly et al., 1980) and decreased ovulation; (Shackell et al., 1993). In one New Zealand study, mixed aged ewes grazed for 12 wk on

Grasslands Pawera red clover before breeding had more returns to service (35 vs. 14%) and had a higher percentage of barren individuals (18 vs 59%) than ewes in a control group grazing ryegrass (Shackell and Kelly, 1984). The ewes on red clover had higher live weight gains (58.0 vs. 54.7 kg). Long terms studies have also shown that ewes grazing pastures containing 60% Grasslands Pawera red clover for about 6 months per year for four years had a higher incidence of vaginal abnormalities than a control group on 0% red clover (Shackell et al., 1993).

Numerous studies have shown that formononetin content of red clover is under genetic control and varies among and within cultivars (Frensis et al., 1981; Agaev et al., 1984; Kallela, 1989; Kallela et al., 1988; Burda et al., 1991). Programs of selection for cultivars with low formononetin content have been conducted in Britain, New Zealand, and Russia. In Britain, levels were reduced from 0.60% to 0.12% in the cultivar Norseman (Gosden et al., 1984). Previous research had suggested that levels lower than 0.20% would be needed to avoid adverse oestrogenic effects (Moseley et al., 1984). In New Zealand, G27, a 6th generation of selection for low formononetin within Grasslands Pawera was developed. Formononetin level ranged between 0.59 and 0.34% in two trials using this selection (Anwar et al., 1993). Additional selection for even lower oestrogenic level in G27 has been conducted. Grasslands Colenso, a cultivar developed in New Zealand for improved winter-season growth is reported to have a formononetin level only 55% of that of Grasslands Pawera (Claydon et al., 1993).

Stage of plant development, plant part, and environment can affect formononetin content of red clover. Content in cv. Hungaropoly was 0.56% when harvested in early May, but decreased to 0.35% when harvested in mid-June (McMurray et al., 1986). In regrowth, youngest material had the highest formononetin content (0.68%). Much of the decrease with maturity appears to be due to an increase in stem content (stems having lower levels than leaves). A lower temperature regime in a pot study (17/13 °C day/night vs 23/15 °C) resulted in higher formononetin content (McMurray et al., 1986). Phosphate fertilizer addition at 96 kg/ha decreased formononetin content compared to red clover grown on low P soil. Research from Sweden suggests that formononetin levels in second cut forage was often higher than in spring cut at similar maturity stages (Kallela et al., 1987). Formononetin generally does not affect larger animals because of animal size, thus no research was found with beef animals. Biochanin A, genistein, and daidzein also have been identified from red clover but generally have not been implicated to have negative impacts on animal fertility.

8. Slaframine

Slaframine is a piperidine alkaloid metabolite produced when animals consume red clover contaminated with the fungus *Rhizoctonia leguminicola*

Gough & Elliott, the causative agent of black spot disease of red clover (Essig, 1985). Animals consuming red clover hay contaminated with this fungus may exhibit excessive salivation syndrome. This syndrome is of primary concern when red clover hay is fed to horses. The fungus is present in most clover fields and no genetic resistance is available. It is generally not a problem on first cutting hay. Making second and later cuttings at the early (1/10) bloom stage will decrease the time for fungal infection and likelihood of problems. Growing grass-clover mixtures will also usually reduce problems with this syndrome. Research from Canada has shown that consumption of red clover contaminated with 7 mg/g slaframine increased salivary flow and tended to increase rumen fluid dilution rate and the acetate to propionate ratio in wethers (Froetschel et al., 1984). A subcutaneous administration of purified slaframine at 4.4 or 11.1 mg/kq increased salivary flow rate over 300% in steers. The problem can usually be corrected by removing the contaminated hay.

References

Agaev, F. N., G.V. Samorodova-Bianki and N.A. Mukhina. 1984. Estimation of oestrogenic isoflavone content in red clover samples. Sel'skokhozyaistvennaya-Biologiya. No. 2 pp. 31–33.

Allen, V.G., J.P. Fontenot and D.R. Notter. 1992a. Forage systems for beef production from conception to slaughter: II. Stocker systems. *J. Anim. Sci.* 70:588–596.

Allen, V.G., J.P. Fontenot, D.R. Notter and R.C.J. Hammes. 1992b. Forage systems for beef production from conception to slaughter: I. Cow-calf production. *J. Anim. Sci.* 70:576–597.

Aman, P. and E. Nordkvist. 1983. Chemical composition and *in-vitro* degradability of major chemical constituents of red clover harvested at different stages of maturity. *J. Food and Agric. Sci.* 34:1185–1189.

Aman, P. 1985. Chemical composition and in vitro degradability of major chemical constituents in botanical fractions of red clover harvested at different stages of maturity. *J. Food and Agric. Sci.* 36:775–780.

Anderson, S. 1986. The effect of cutting date and nitrogen fertilizer on overwintering of pastures. *Nordish Jordbrugsforsking.* 68:571.

Anwar, A., R.G. Keogh and M.F. McDonald. 1993. Oestrus, ovarian activity and embryo survival in ewes grazing high and low oestrogenic red clover. *Proc. of the N. Z. Soc. of Animal Prod.* 53:303–305.

Baker, E.L., M. Collins and N.L. Taylor. 1993. Developmental patterns and hay drying effects of red clover stem trichomes. pp. 235–239. In *Proc. Americ. Forage Grassl. Council.* 29–31 March, Des Moines, IA.

Baligar, V.C., R.J. Wright, T.B. Kinraide, C.D. Foy and J.H. Elgin Jr. 1987. Aluminum effects of growth mineral uptake and efficiency ratios in red clover cultivars. *Agron. J.* 79:1038–1044.

Barnett, F.L. and G.L. Posler. 1983. Performance of cool-season perennial grasses in pure stands and in mixtures with legumes. *Agron. J.* 75:582–586.

Bawolski, S. and E. Gawel. 1985. Comparison of the yields of several red clover and white clover cultivar and their mixtures with grasses. *Pamietnik-Pulawski.* 85:129–140.

Binderova, A. 1989. Reaction of some varieties and new selections of red clover to harvesting dates. Sbornik Vedeckych Praci–Oseva, Vyzkum Slechteni a Semenarstvi, koncern Praha, Koncernovy Vyzkumny a Slechtitelsky Ustav Picninarsky. *Troubsko u Brna.* 11:153–163.

Bojas, Z. 1988. Use of herbicides for the establishment of clover stands. *Sbornik UVTIZ, Ochrana Rostlin.* 24:141–146.

Bowley, S.R. and S.M. Lackie. 1989. Genetics of nonglandular stem trichomes in red clover (*Trifolium pratense* L.) *J. Hered.* 80:472–474.

Bowman, J.G.P. and J.L. Firkins. 1993. Effects of forage species and particle size on bacterial cellulolytic activity and colonization in situ. *Amer. Soc. Animal Sci.* pp. 1623–1633.

Brookes, I.M. and G.F. Wilson. 1983. Milk production by dairy cows grazing 2 red clover cultivars with and without sodium supplementation. N. Z. *J. Exp. Agric.* 11:137–140.

Brougham, R.W. and G.P. Cosgrove. 1985. Pastures and pasture management for high animal production in New Zealand. *Proc. Inter. Grassl. Cong.* 15:1134 Kyoto, Japan.

Buck, D.G., R.D.H. Cohen and D.A. Christensen. 1989. Effects of various plant growth regulators on the nutritive value and yield of some grass species, red clover and grass-legume mixtures. *Canad. J. Plant Sci.* 69:465–480.

Burda, S., P.M. Gorski, M. Jurzysta and M. Ploszynski. 1991. The formononetin content in some Polish di- and tetraploid red clover cultivars. *ACTA Agrobotanica.* 44:81–85.

Bush, L. and H. Burton. 1994. Intrinsic chemical factors in forage quality. pp. 367–405. In G.C. Fahey, Jr. (ed.), *Forage Quality, Evaluation, and Utilization.* ASA-CSSA-SSSA Madison, WI.

Buxton, D.R. and J.R. Russell. 1988. Lignin constituents and cell-wall digestibility of grass and legume stems. *Crop Sci.* 28:553–558.

Buxton, D.R., J.S. Hornstein, W.F. Wedin and G.C. Marten. 1985. Forage quality in stratified canopies of alfalfa, birdsfoot trefoil and red clover. *Crop Sci.* 25:273–279.

Buxton, D.R. and J.S. Hornstein. 1986. Cell wall concentration and components in stratified canopies of alfalfa, birdsfoot trefoil, and red clover. *Crop Sci.* 26:180–184.

Campbell, B.D. and H.T. Kunelius. 1984. Performance of overdrilled red clover *Trifolium pratense* cultivar Grasslands-Pawera with different sowing rates and initial grazing managements. *N. Z. J. Exp. Agric.* 12:71–82.

Chlevickas, J., T. Chlevickiene, D. Skebiene and D. Jankeviciene. 1990. Quality of silages made from legumes and grasses with biological and chemical additives. Soil-grassland-animal relationships. Proceedings of (13) general meeting of the European Grassland Federation, Banska Bystrica, Czechoslovakia, June 25–29, 1990. pp. 101–104.

Choo, T.M. and M. Suzuki. 1983a. Effect of plant age on cold hardiness in red clover. *Forage Notes.* 27:28.

Choo, T.M. and M. Suzuki. 1983b. Cold hardiness of six red clover varieties. *Forage Notes.* 27:26–27.

Clark, E.A., S.V. Crump and Z.P. Kondra. 1989. Mechanical and chemical conditioning of hays. *Canad. J. Plant Sci.* 69:133–141.

Claydon, R.B., J.E. Miller and L.B. Anderson. 1993. Breeding a winter growing red clover - cv. Grasslands Colenso. *N. Z. J. Agric. Res.* 36:297–300.

Collins, M. 1983. Changes in composition of alfalfa, red clover and birdsfoot trefoil during autumn. *Agron. J.* 75:287–291.

Collins, M. 1985a. Wetting effects on the yield and quality of legume and legume-grass hays. *Agron. J.* 77:936–941.

Collins, M. 1985b. Leaching effects on composition and digestibility of leaf and stem of alfalfa and red clover. *Proc. Intern. Grassl. Cong.* 15:995–997. Kyoto, Japan.

Collins, M. 1989. Conditioning effects and field variation in dry matter concentration of alfalfa hay. *Proc. Intern. Grassl. Cong.* 16:1005–1006. Nice, France.

Collins, M. and T.H. Taylor. 1984. Quality changes of late summer and autumn produced alfalfa and red clover. *Agron. J.* 76:409–415.

Coulman, B.E. and A. Kielly. 1988. The effect of seeding rate and cutting frequency on the yield and persistence of red clover. *Canad. J. Plant Sci.* 68:733–742.

Decker, A.M., R.F. Dudley, L.R. Vough, M.I. Spicknall and T.H. Miller. 1986. Band vs. broadcast paraquat and alternative row species seeding in no-till pasture renovation. *Amer. For. Grassl. Council. Proc.* Lexington, KY. pp. 254–261.

Eraso, F. and R.D. Hartley. 1990. Monomeric and dimeric phenolic constituents of plant cell walls - possible factors influencing wall biodegradability. *J. Food and Agric. Sci.* 51:163–170.

Essig, H.W. 1985. Quality and antiquality components. In N.L. Taylor (ed.) *Clover Science and Technology.* ASA-CSSA- SSSA Mono. 25:309–324. Madison, WI.

Fay, M.F. and P.J. Dale. 1984. Legume bloat and tannins in *Trifolium* species. In *Proc. British Grasslands Society Symp.*, Berkshire College of Agriculture, Burchett's Green, Maidenhead, England pp. 21–23.

Fordonski, G., K. Kowalik, I. Klicka and S. Goral. 1988. Effect of sowing method on yield and fodder quality in one-cut and two-cut red clover varieties. *Biuletyn Instytutu Hodowli i Akimatyzacji Roslin.* 168:87–96.

Frame, J., R.D. Harkess and I.V. Hunt. 1985. Effect of seed rate of red clover and of companion timothy or tall fescue on herbage production. *Grass Forage Sci.* 40:459–465.

Frensis, K.M., N.A. Mukhina and C.M. Francis. 1981. Content of the isoflavone formononetin and yield of dry matter in forms of red clover. *Trudy-po-Prikladnoi-Botanike,-Genetike-i-Selektsii.* 71:138–141.

Freudenberger, D.O., C.J. Burns, K. Toyokawa and T.N. Barry. 1994. Digestion and rumen metabolism of red clover and perennial ryegrass/white clover forages by red deer. *Agric. Sci. J.* 122:115–120.

Froetschel, M.A., W.J. Croom, W.M. Hagler, H.P. Broquist and R. Gaskins. 1984. Effects of slaframine on salivary flow and rumen function. *J. Conad. Animal Sci.* 64:64–65.

Gosden, A.F., W.E. Davies, L. Hughes and R. Jones. 1984. Breeding for reduced formononetin in red clover. In *Proc. British Grasslands Society Symp.* 21–23 Feb 1984, Berkshire College of Agriculture, Burchett's Green, Maidenhead, England. pp. 21–23.

Guillard, K. and D.W. Allinson. 1987. Establishment of legumes using no-till seedings with and without herbicides. *Appl. Agric. Res.* 1:281–288.

Guy, P. 1989. Multisite trials of red clover-grass mixtures. *Forages.* 117:29–48.

Habovstiak, J. 1989. Effect of different liming rates on meadow clover (*Trifolium pratense* L.) yields in mountainous regions of Slovakia. *Pol'nohospodarstvo.* 35:519–527.

Haggar, R.J. and D.W. Koch. 1983. Slot seeding investigations 3. The productivity of slot seeded red clover *Trifolium pratense* cultivar Hungaropoly compared with all grass swards receiving nitrogen. *Grass Forage Sci.* 38:45–54.

Hall, M.H. and J.W. Eckert. 1992. Seeding year harvest management of red clover. *J. Prod. Agric.* 5:52–57.

Halling, M.A. 1988. Influence of autumn cutting time and weather on growth potential and growth of timothy (*Phleum pratense* L.) and red clover (*Trifolium pratense* L.). *Crop Prod. Sci.* 1:83.

Hickey, M.J. and A.J. Harris. 1989. Comparative performance of three early-flowering red clovers under sheep grazing in Southland New Zealand. *N. Z. J. Agric. Res.* 32:327–332.

Hoden, A., B. Marquis and F.X. de la Foye. 1987. Maize and red clover silages for dairy cows. *Bulletin Technique Centre de Recherches Zootechniques et Veterinaires de Theix.* 67:33–37.

Hong, B.J., Y.G. Goh, S.J. Ohh and Y.C. Rhee. 1989. Effect of chemical conditioning on the drying rate and ruminal nutrient digestion of red clover. L. *Korean J. Animal Nutrition Feed.* pp. 201–207.

Hrabe, F. and E. Halva. 1989. Effect of nutrition and supplementary irrigation on energy production and yield formation in red clover (*Trifolium pratense* L.). *Rostlinna Vyroba.* 7:733–740.

Hunt, W.F. and J.M. Hay. 1990. A photographic technique for accessing the pasture species performance of grazing animals. *Proc. N. Z. Grassl. Assn.* 51:181–186.

Igarashi, K. and T. Yasui. 1989. Changes in free proline content of red clover (*Trifolium pratense* L.) during forced air drying. *J. Japan Soc. Grass Sci.* 35:141–145.

Igarashi, K., H. Kamiyama, M. Tezuka and T. Yasui. 1985. Changes in amino-acid composition during sun-curing of leaves and stems of red clover trifolium-pratense. *Japan. J. Zootechnical Sci.* pp. 566–570.

Jezikova, O. and M. Michalec. 1988. Additional sowing of legumes directly in the sward for the purpose of improving grass stands and their quality. *Ved Pr Vysk Ustavu Luk Pasienkov Banskej Bystrici.* 19:217–228.

Johnson, T.R., J.W. Thomas, C.A. Rotz and M.B. Tesar. 1984. Drying rate of cut forages after spray treatments to hasten drying. *J. Dairy Sci.* pp. 1745–1751.

Kallela, K. 1989. Plant oestrogens in red clover; occurrence and significance in Finland. *Nordisk-Jordbrugsforskning.* 71:108.

Kallela, K., I. Saastamoinen, E. Huokuna and H. Hakkola. 1988. Variation in plant oestrogen content between certain red clover cultivars in northern and southern Finland. *Suomen-Elainlaakarilehti.* 94:287–291.

Kallela, K., I. Saastamoinen and E. Huokuna. 1987. Variations in the content of plant oestrogens in red clover-timothy grass during the growing season. *Acta Veterinaria Scandinavica.* 28:255-262.

Kalmbacher, R.S. and F.G. Martin. 1982. Evaluation of clover seeded at three dates in grazed or burned bahiagrass. *Proc. Soil and Crop Sci. Soc. FL.* 41:84–87.

Kawakami, K. 1982. Studies on hay conditioning by a pair of rollers (II). Comparison between crimper and crusher type rollers. *J. Soc. Agric. Machinery, Japan.* 44:447–454.

Kelly, R.W., G.H. Shackell and A.J. Allison. 1980. Reproductive performance of ewes grazing red clover (Grasslands Pawera) or white clover-grass pasture at mating. *N. Z. J. Exp. Agric.* 8:87–91.

Kelly, R.W. and G.H. Shackell. 1982. Ovulation and oestrous responses of high and low fecundity ewes to ingestion of isoflavone-rich pasture. *Proc. N. Z. Soc. Animal Prod.* 42:29–31.

Kendall, W.A. 1966. Factors affecting foams with forage legumes. *Crop Sci.* 6:487–489.

Kharkov, G.D. and V.A. Starikov. 1985. Yield of red clover and lucerne in relation to techniques for cultivation of cover crop. Nauch. issled. v kormoproizvodstve Volgo-Vyatskoi zony. Kirov, USSR. pp. 11–16.

Kiso, S. and K. Kikuchi. 1988. Optimizing the application rates of nitrogen fertilizer on the basis of legume ratio in timothy (*Phleum pratense* L.) dominant meadows. *J. Japan. Soc. Grassl. Sci.* 34:169–177.

Kondo, T., K. Mizuno and T. Kato. 1986. Chemical and physical properties of forage plant lignin. I. Differential effect of acid detergent on grass and legume lignin. *Bulletin of National Grass. Res. Inst., Japan.* pp. 55–61.

Kuhbauch, W. and J. Bestajovsky. 1989. Distribution of lignin in stems of lucerne and red clover. *Wirtschaftseigene-Futter.* 35:19–28.

Laidlaw, A.S. 1983. Grazing by sheep and the distribution of species through the canopy of a red clover *Trifolium pratense* cultivar Hugaropoly perennial rye grass *Lolium perenne* cultivar Barlenna sward. *Grass Forage Sci.* 38:317–322.

Lang, B.J. 1985. The effect of fertility and harvest management on yield, persistence, reserve carbohydrate storage, regrowth, and quality of alfalfa and red clover. *M.S. Thesis, University of Wisc.* 189 pp.

Lee, I.D, J. Myung, W.S. Song and Y.K. Chun. 1987. Study on the use of orchardgrass-red clover mixtures. I. Effect of plant composition on intake, digestibility and preference by Korean native goats. *J. Korean Soc. Grass. Sci.* 7:31–36.

Lemieux, C., A.K. Watson and J.M. Deschenes. 1987. Effect of barley, red clover and weeds on yield and forage quality of timothy during establishment and subsequent production years. *Canad. J. Plant Sci.* 67:1019–1033.

Lesak, J. and J. Sverakova. 1988. Results of agrotechnical trials with clover/grass mixtures. *Rostinna-Vyroba.* 34:1221–1229.

Liu, Y.G., Z.C. Xu, W.Y. Huang, S.X. Zhang and S.Q. Yang. 1991. Digestibility and feeding value of red clovers for pigs. *Chinese J. Animal Sci.* 27:6–8.

Lowe, L.B., G.J. Ball, V.R. Carruthers, R.C. Dobos, G.A. Lynch, P.J. Moate, P.R. Poole and S.C. Valentine. 1991. Monensin controlled-release intraruminal capsule for control of bloat in pastured dairy cows. *Aust. Vet. J.* 68:17–20.

McKenny, D.J., S.W. Wang, C.F. Drury and W.I. Findlay. 1993. Dentrification and mineralization in soil amended with legume, grass, and corn residues. *Soil Sci. Soc. Am. J.* 57:1013–1020.

McMurray, C.H., A.S. Laidlaw and M. McElroy. 1986. The effect of plant development and environment on formononetin concentration in red clover (*Trifolium pratense* L.). *J. Sci. Food Agric.* 37:333–340.

Mengel, K. and D. Steffens. 1985. Potassium uptake of rye-grass (*Lolium perenne*) and red clover (*Trifolium pratense*) as related to root parameters. *Biol. and Fert. of Soils.* 1:53–58.

Mika, V. and C. Paul. 1988. Intake of some Czechoslovak varieties of red clover and grasses by cattle. *Rostlinna-Vyroba.* 34:553–560.

Mitchell, R.L., J.B. McLaren and H.A. Fribourg. 1986. Forage growth, consumption, and performance of steers grazing bermudagrass and fescue mixtures. *Agron. J.* 78:675–680.

Moran, J.B. and W.N. Wamungai. 1992. Maize silage for the pasture-fed dairy cow: 4. A comparison between grazed annual pastures and feedlotting in the winter. *Aust. J. Exp. Agric.* 32:301-309.

Moseley, G., R. Jones and G.H. Shackell. 1984. The effects of red clover diets on the reproductive function of the ewe. In *Proc. British Grasslands Society Symp.* 21–23 Feb. 1984, Berkshire College of Agriculture, Burchett's Green, Maidenhead, England. pp. 21–23.

Mueller, M.M. 1988. The fate of clover derived nitrogen-15 during decomposition under field conditions effects of liming and fertilization. *Plant and Soil.* 111:121–126.

Narasimhalu, P. and T.M. Choo. 1983. Effect of potassium carbonate on drying of red clover under humid field conditions. *Zeitschrift für Acker und Pflangenbau.* 152:159–163.

Niezen, A.H., T.N. Barry, J. Hodgson, P.R. Wilson and C.W. Holmes. 1991. The effect of three allowances of red clover on red deer fawns growth and liveweight change in lactating hinds. *Proc. N. Z. Soc. Anim. Prod.* 52:185–187.

Niezen, J.H., T.N. Barry, J. Hodgson, P.R. Wilson, A.M. Ataja, W.J. Parker and C.W. Holmes. 1993. Growth responses in red deer calves and hinds grazing red clover, chicory or perennial ryegrass/white clover swards during lactation. *N.Z. J. of Agric. Sci.* 121:255–263.

Oyen, J. 1991. Methods of establishing grass-clover swards. *Norway J. Agric. Sci.* 5:301–306.

Peaslee, D.E. and N.L. Taylor. 1989. Red Clover. pp. 447–458. In D.L. Plucknett and H.B. Sprague (eds.) *Detecting Mineral Nutrient Deficiencies in Tropical and Temperate Crops.* Westview Press, Inc. Boulder, CO.

Pederson, G.A. and G.E. Brink. 1991. Productivity and quality of annual and perennial clover-tall fescue mixtures. *Agron. J.* 83:694–699.

Pedraza, G.C., A.I. Tames and H.H. Olguin. 1988. Milk production from direct grazing of birdsfoot trefoil, lucerne and red clover. *Agricultura Tecnica.* 48:97–101.

Peterson, P.R., C.C. Sheaffer and M.H. Hall. 1992. Drought effects on perennial forage legume yield and quality. *Agron. J.* 84:774–779.

Ray, R.J.M. and W.F. Hunt. 1989. Competition from associated species on white and red clover in grazed swards. Persistence of forage legumes. *Proc. Trilateral Workshop* 18–22 July, 1988, Honolulu, HI. pp. 311–326. ASA Madison, WI.

Sawai, A., M. Gau and S. Ueda. 1986. Difference in root system among growth types of red clover. *J. Japan. Soc. Grassl. Sci.* 32:164–166.

Semiadi, G., T.N. Barry, P.R. Wilson, J. Hodgson and R.W. Purchas. 1992. Growth and venison production from red deer grazing either pure red clover or perennial ryegrass/white clover pasture. *Proc. N. Z. Soc. Animal Prod.* 52:81–83.

Shackell, G.H. and R.W. Kelly. 1984. Residual effects of short-term grazing of red clover cultivar Grasslands-Pawera pastures on the reproductive performance of ewes. *N. Z. J. Exp. Agric.* 12:113–118.

Shackell, G.H., J.G. Wylie and R.W. Kelly. 1993. Effects of prolonged exposure of ewes to oestrogenic pasture: 2. Occurrence of abnormalities of the external genitalia and altered mating performance. *N. Z. J. Agric. Res.* 36:459–464.

Sheldrich, R.D., R.H. Lavender and V.J. Tewson. 1986. The effect of frequency of defoliation data of first cut and heading date of a perennial ryegrass companion on the yield, quality and persistence of diploid and tetraploid broad red clover *Trifolium pratense*. *Grass Forage Sci.* 41:137–150.

Shuppenies, R. 1988. Comparison of grass and grass/clover on a deep ploughed sandy site. *Feldwirtschaft.* 29:83–84.

Siebald, S.E., M.L. Goic, V. Elizalde, H.F. and K.M. Matzner. 1989. Evaluation of red clover-ryegrass silages according to cutting date and sodium formate treatment. *Agricultura Tecnica Santiago.* 49:352–356.

Silva, G.M., T.H. Lopez and V.A. Pascual. 1988. The influence of cutting height on the nutritive value of red clover (*Trifolium pratense*) cultivar Quinequeli. *Agricultura Tecnica.* 48:188–196.

Sineiro, F., K. Osora and N. Diaz. 1984. Development of systems of production with beef cows in the hill country of Galicia (Spain). Hill Land Symposium. Dublin, Irish Republic. pp. 300–329.

Smith, R.R., N.L. Taylor and S.R. Bowley. 1985. Red clover. In N.L. Taylor (ed.) *Clover Science and Technology.* ASA-CSSA-SSSA Mono 27:457–470. Madison, WI.

Staley, T.E. 1993. Growth of perennial forage legumes in acidic soils of the Appalachian hill-lands after liming. *J. Plant Nutr.* 16:2577–2590.

Tagaev, O.A., Y. A. Pazderskii, B.D. Popovich, F.I. Vrydnyk and I.I. Moiseev. 1985. Formic acid - a preservative for green feeds. *Khimiya v Sel'skom Khozyaistve.* 23:58–59.

Tanaka, A., K. Hitsuda and Y. Tsuchihashi. 1984. Tolerance to low pH and low available phosphorus of various field and forage crops. *Soil Sci. Plant Nutr.* 30:39–49.

Takeda, Y., Y. Sagae, H. Yomazahi and H. Mekito. 1989. Improvement of timothy (*Phleum pratense* L.) dominant swards by sod seeding of red clover (*Trifolium pratense* L.) I. Suppression of timothy with paraquat application and seed-bed preparation by surface cultivation. *J. Jap. Soc. Grassl. Sci.* 35:212–219.

Thomas, C., K. Aston and S.R. Daley. 1985. Milk production from silage. 3. A comparison of red clover with grass silage. *Animal Prod.* 41:23–31.

Trimmer, M.C. and D.L. Linscott. 1986. Herbicide effects on water use by grass sod and red clover *Trifolium pratense* establishment. *Weed Sci.* 34:48–51.

Wilman, D. and M.A.K. Altimimi. 1984. The *in-vitro* digestibility and chemical composition of plant parts in white clover *Trifolium repens* cultivar grasslands Huia, red clover *Trifolium pratense* cultivar norseman, and lucerne *Medicago sativa* cultivar Sabilt during primary growth. *J. Food and Agric. Sci.* 35:133–138.

6. Fungal Diseases

In the last decade, most research on diseases of red clover has concentrated on crown rot caused by *Sclerotinia trifoliorum* Erikss and the root rot complex, caused by a large number of pathogens but associated most often with species of *Fusarium*. Much of this research is being conducted in Eastern Europe, and New Zealand. A survey of diseases other than viruses and nematodes in the United States and Canada, showed about 18 diseases affecting red clover (Leath, 1989). Many of these were indicated to be of minor importance or only occasionally severe. The major diseases are included here (Table 6.1) for historical purposes. However, this review will concentrate on recent research primarily on the crown and root-rot complexes. See Leath (1985) for discussion of earlier literature.

1. The anthracnoses

Southern anthracnose caused by *Colletotrichum trifolii* B & E was one of the first diseases for which resistance was obtained. The disease occurs in the warmer region of North America, Kenya and South Africa. Lesions occur on leaves, stems, and crowns causing a typical "shepherd's crook" of the petioles, and eventually result in death of the infected plant. Resistance to the disease is apparently conditioned by one or few genes but genetics of resistance has not been completely elucidated. Cultivars that carry resistance include Kenland and Kenstar, and several related commercial cultivars. Probably as a result of this resistance, serious epiphytotics of this disease have not occurred in the last 20 or so years.

Northern anthracnose caused by *Kabatiella caulivora* (Kirch.) Karak. has symptoms similar to southern anthracnose except that death of plants usually does not occur directly (Fig. 6.1). However, quality of forage is reduced, and stand persistence is lessened due to the interaction with other diseases, especially root rots. The disease occurs mostly in the cooler regions of North America, Europe and Asia. Resistance to the disease, controlled by two or three dominant genes, is present in the cultivars Lakeland, Arlington, and Marathon, and related commercial cultivars.

Table 6.1

Major diseases that primarily attack red clover (other than nematodes and virus diseases).

Common name	Causal agent	Control
Black patch	*Rhizoctonia leguminicola* Gough & Elliott	Cultural
Target spot	*Stemphylium sarcinaeforme* (Cov.) Wiltsh.	Resistance
Powdery mildew	*Erysiphe polygonic* D C em Salm	Resistance
Rust	*Uromyces trifolii* (Hedw. f. ex DC)	Resistance
Northern anthracnose	*Kabatiella caulivora* (Kirchn) Karak.	Resistance
Southern anthracnose	*Colletotrichum trifolii* Bain & Essary	Resistance
Southern root and stolon rot	*Sclerotium bataticola* (Taub.) and *S. rolfsii* Sacc.	None
Crown rot	*Sclerotinia trifoliorum* Erikss	Resistance
Root rots	*Fusarium* spp. among others	None

Recently, research has been conducted to combine resistance to both anthracnose diseases in several commercial cultivars, and in the Kenstar cultivar (Taylor et al., 1990). Phenotypic recurrent selection was effective in uncovering latent variation for resistance to *K. caulivora*. Mean disease severity index was reduced by 36% by 6 cycles of selection. The resultant synthetic, although not completely resistant, was judged to have sufficient resistance to perform adequately under field conditions.

Breeding for resistance to northern anthracnose is also being conducted in the Volga - Vyalka area of European Russia (Nikiforova and Suslova, 1986). Several cultivars with anthracnose resistance have been approved for cultivation. Also, in Europe, earlier research has indicated that tetraploid cultivars have greater resistance than diploid cultivars.

2. Mildew

The most serious mildew on red clover is powdery mildew caused by *Erysiphe polygoni* DC. The disease is prevalent wherever red clover is grown, but is not as serious in the central part of the clover belt of the U.S. as it is in the northern part, or in southern U.S. where red clover is an annual or biennial. The growth of mycelium and powdery appearance of conidia give a conspicuous white or light grey cast to the leaves (Fig. 6.2) which, if infection is severe, turn yellow to brown. Quality of forage is apparently decreased but no evidence on yield has been obtained. The disease is easily controlled by resistance genes which are dominant, and monogenic for most races of the mildew. Selection

Fig. 6. 1. Northern anthracnose symptoms.

Fig. 6. 2. Powdery mildew symptoms.

for seedling resistance under greenhouse conditions usually confers mature plant resistance under field conditions.

Recently, downy mildew caused by *Peronospora trifolii-pratensis* has been found on red clover in Sweden, but causes little damage (Gustavsson, 1988).

3. Rust

This disease, caused by *Uromyces trifolii* var *fallens* produces pustules on leaves, stems and petioles. If the disease is severe, death and loss of leaves may occur. Rust occurs widely throughout humid and semi-humid regions primarily in late summer or early autumn. In the U.S., it may be particularly severe in the northeastern states. The disease has not been deemed to be sufficiently severe to warrant development of resistant cultivars, even though simply inherited resistance is undoubtedly available among plants of northern U.S. cultivars.

4. Target spot or Stemphylium leaf spot

This is probably the most important of several leaf spots diseases that affect red clover. Causal agent is *Stemphylium sarcinaeforme* (Cav.) Wiltshire. It is prevalent in most humid regions. Symptoms are small to large irregular dark brown, sunken spots on leaves, stems, and petioles. Several sources of resistance are available, and breeding for resistance has been conducted at USDA, ARS at the University of Wisconsin (Smith and Kretschmer, 1989). Four cycles of phenotypic selection increased resistance by about 24%, about 6% per cycle. Some resistant materials probably have been included in northern USA cultivars.

5. *Aphanomyces euteiches* Drechs

This is a fungal soil-borne pathogen of which one isolate is highly virulent to red clover (Holub, 1988). Three cycles of seedling selection for resistance to *A. euteiches* resulted in significant improvement (Tofte et al., 1991). Narrow sense heritability in cycle 3 was 51% suggesting that further improvement was possible. To determine the inheritance of resistance, resistant plants from the cycle 2 population, and susceptible plants from the base population were crossed and progenies were evaluated. Resistance appeared to be inherited as a quantitative trait. However, the plants used as resistant in cycle 2 were not completely resistant suggesting that further study may be necessary.

6. Sclerotinia crown rot

The causal agent of this disease, *Sclerotinia trifoliorum* Eriks. is responsible for widespread losses of red clover due to death of plants in the spring season. The disease is very severe in regions of Europe and Japan that have mild winters and heavy snows. In the United States, the disease is sporadic and is most prevalent in the southern clover belt including Virginia, Kentucky and Tennessee. Infection first occurs in late autumn when sclerotia germinate to produce apothecia from which ascopores are discharged (Fig. 6.3). The spores cause a pepper spotting of leaves which drop off and are overrun by white mycelial growth. Continued progress of infections in the winter and spring results in a soft rot, often under snow cover, resulting in a dead plant.

Fig. 6. 3. Apothecia of *Sclerotinia trifolium.*

Fig. 6. 4. Severe root rotting in crown of red clover.

Black sclerotia, the resting stage of the fungus, may be found around the base of dead plants (Scott and Evans, 1984).

Several practices may be recommended to lessen the severity of crown rot, but complete control may not be achieved if the environmental conditions are conducive to development of the disease. Firstly, in the U.S. at least, sowing red clover in the spring rather than in the fall may prevent the disease, or lessen its severity. Also, inasmuch as *sclerotinia* may be carried with the seed, it is best to not use seed from fields known to be heavily infected with *Sclerotia* crown rot (Scott and Evans, 1984). Biological control of crown rot by *Coniothyrium minitans*, a known parasite of sclerotia, has been shown to be effective when red clover is grown in monoculture. The parasite reduced apothecial formation by 88% in monoculture red clover as compared to red clover grown in a rotation (Pfeffer and Luth, 1990). The authors concluded that it is not necessary to select for resistance to *S. trifoliorum* under monoculture conditions. Application of the chemical vinclozolin during sporulation of the fungus (mid-Oct. to mid-Dec. in Germany) has been reported to provide 80 to 100% control (Sauer, 1984). The chemicals carbendazin and iprodione were the most effective fungicides tested for control of *S. trifoliorum* under growth chamber conditions (Raynal and Picard, 1985). Unless the disease is very severe, it is doubtful that the cost of fungicides can be justified.

It has been long known that breeding for resistance to *S. trifoliorum* was possible. Firstly, cultivars differ in natural resistance to the fungus. The cultivar Kenland from the southern part of the clover belt in the U.S. possesses a low level of resistance although no breeding has been conducted. The tetraploid form of Kenstar developed in the southern part of the U.S. clover belt has been one of the most resistant cultivars in laboratory and field tests (Arseniuk, 1989). Likewise, many Swedish cultivars exhibit natural resistance or tolerance to the disease. Secondly, tetraploid cultivars are generally, but not always, more resistant than diploids. It appeared that susceptible tetraploids were often derived from susceptible diploids, and resistant tetraploids from resistant diploids. Based on these and similar data, breeding for resistance has been of paramount importance to breeders and plant pathologists for many years. Screening for resistance may be achieved either by planting material for selection in land known to contain the disease, or by using artificial inoculation techniques in either field or controlled environments. Numerous techniques have been described that will rank cultivars between the extremes of susceptibility and resistance. However, most of these tests are notably unreliable and unsatisfactory for selecting individual resistant plants. Fungal isolates, if uncontrolled, usually will destroy all plants, both susceptible and resistant. Although it has been shown that infection of plants may occur below ground from mycelium growing from sclerotia, primary inoculum is from ascospores. A laboratory method of growing the fungus on autoclaved potato slices has been devised, which produces sclerotia that later germinate to produce apothecia (Raynal and Picard, 1985). Ascospores were collected by desiccating inverted apothecia in Petri dishes (Raynal, 1983). Red clover plants, 3 months old, were sprayed with an aqueous suspension of ascospores, and scored for infection at appropriate later dates. The assessment of cultivar resistance under these conditions was well correlated with field resistance. Reaction of red clover cultivars inoculated either with mycelium or with ascospores was similar with a few exceptions according to Marum et al. (1994). The authors suggested that ascospore inoculation was superior because mycelium-inoculated derived germplasm has not proven to express acceptable resistance under field conditions. If selection of plants by the ascospore procedure adequately confers field resistance, a major step in increasing the reliability of red clover stand maintenance will have been achieved.

7. Root rot

In the last 10 to 15 years, much literature deals with root-rot diseases of red clover. The primary causal agents are fungi in the genus *Fusarium*. Species most often mentioned are *F. oxysporum* Schlect, *F. solani* Mart (Appel), *F. avenaceum*, or *F. roseum* avenaceum (Fig. 6.4). Another species mentioned occasionally is *F. culmorum*. Other genera implicated in root rot of red clover include *Pseudomonas* (Leath et al., 1989), in Pennsylvania, USA, *Thielovioppsis, Cylindrocladium*, and *Verticillium* (Skipp et al., 1986) in New Zealand, and *Cylindrocarpon*, and *Pythium* (Wanson and Maraite, 1984) in Belgium. Interactions of these and other organisms such as nematodes and insects are involved in the root-rot complex (Jin et al., 1991).

Control of root rot in red clover is difficult. Because root-rotting organisms tend to build up in soils when red clover is grown continually, an obvious control measure is to use crop rotations in which red clover is less frequently sown, or to change to a crop that is not susceptible to the same root-rot organisms as is red clover. Other possible controls include use of methyl bromide or other soil fumigants, or drenches. However, the cost of these treatments eliminates them from consideration except for special purposes.

Resistance breeding is an obvious method for control of root rot. Various artificial techniques have been used to access and screen for resistance in red clover cultivars. *In vitro* inoculation procedures include those of Nedelnik (1986, 1989) who inoculated plants in the cotyledon stage in sterile test tubes with a mixture of *F. oxysporum* and *F. solani*. Cultivars were differentiated according to their resistance. A similar test was used to measure the frequency and intensity of seed infection on plants grown on a nutrient medium in test tubes. Intensity of infection was highest in seeds of moderately to highly susceptible cultivars (Kovacikova and Kudela, 1989). High correlation of laboratory and field assessments of the resistance of 16 cultivars of red clover indicated that laboratory tests could be used for preliminary screening of breeding materials before tests start under natural conditions (Nedelnik, 1989).

Indirect methods of assessing root-rot resistance *in vitro* have been attempted by a number of investigators. Biological activity of toxic fungal metabolites of *F. roseum* has been shown to be correlated with virulence of the isolates (Stutz et al., 1984). Culture filtrates of *F. roseum* also have been tested on red clover (Blain et al., 1991). Although the extracts were toxic to germinating seeds, they showed only slight host specificity for susceptible red clover seedlings. An examination of rapid methods of diagnosing resistance to *Fusarium* disclosed a high correlation between mean scores for infection in a field test and change in the amount of phenol compounds in response to infection (Paplauskiene, 1988). On the other hand, in another study, no relationship between root amino acid content and level of resistance was found (Kovacikova et al., 1989).

In vivo methods for evaluation and/or screening for resistance to root rots of red clover have been used by Leath et al. (1989) who showed that *Fusarium* and *Pseudomonas* spp. inoculated together into wounded roots of four-week old plants on slant boards caused longer rot lesions and more severe disease symptoms than when inoculated separately. Another *in vivo* method is that of Nedelnik (1988) who dipped rooted cuttings in a suspension of *Fusarium* spp. spores and was able to differentiate resistance in diploid and tetraploid cultivars. A similar method has been used by Rufelt (1985) who placed severed roots in pots in contact with *Fusarium* inoculum. After two cycles of selection, progenies had less root rot than their parents. Anderson and Kristiansson (1989) conducted three cycles of recurrent selection for resistance to an isolate of *Fusarium avenaceum* by the severed root technique. In both the tetraploid cultivar Molly, and the diploid cultivar Essi II, the severity of root rot was decreased as measured by greenhouse tests. The authors indicate that the material must be tested under field conditions to determine if populations have been improved.

Many investigators report the evaluation and screening for root-rot resistance under greenhouse and/or field conditions, either under natural or artificially induced epiphytotics. One method of artificial infection is to scatter *Fusarium*-infected grain prior to sowing red clover seeds either in a greenhouse or field (Putsa and Mezentseva, 1989). Measures also were taken to increase stress such as; later sowing and frequent cutting. Cutting the stand at two-week intervals also may increase root rot severity (Vanco, 1987). In the former Czechoslovakia, two cycles of selection for root rot resistance under field conditions reduced infection of rots caused by *Fusarium* spp., but green matter yield was lower (Uzik, 1983).

It is interesting to note that, under field conditions, factors of general adaptability were often involved with resistance to root rot. The best results of selection are usually found in local cultivars (Nagovitsyna and Gracheva, 1990). Also, a correlation between winterhardiness and resistance to root rots has been reported (Smurygin and Vinnik, 1991). Likewise, in the former Czechoslovakia, the local cultivar Kvarta was found to be more resistant to *Fusarium solani* and *F. oxysporum* than world collection cultivars under field conditions (Kovacikova et al., 1991). Similar data were obtained by Chapurin et al. (1988). These results imply that local adaptation is an important factor in resistance to root rot (See Chapter 10, Persistence).

Another selection method is to dig and select roots for low incidence of rot after one or two seasons in the field (Andersson, 1985). Evaluation of three cycles of such selection under spaced planted conditions indicated that cycle-3 plants were significantly less rotted than plants of cycle 1 or 2. However, Pederson et al. (1980) indicated that such selection tended to reduce root diameter, and Taylor (unpublished) observed a similar relationship. Evaluation under broadcast or drilled conditions will be necessary to validate this procedure.

Little evidence of heritability of root rot resistance exists in the literature although Kovacikova and Jarosik (1993) suggest that the similarity of progeny reaction to *Fusarium* spp. to that of the mother strain indicated that the degree of resistance might be heritable, and that transfer of resistance was possible.

In summary, in the last decade, red clover disease research has centered on crown and root-rot resistance. Most investigators implicate *Fusarium* spp. as the main causes of root rot. Considerable recurrent selection is underway, but it remains to be seen whether resistance to a single fungal genus will confer resistance to root rot over a wide range of conditions. Screening for resistance to all causal agents of the root environment along with general adaptability is likely to show the most promising results.

References

Andersson, B. 1985. Some results of selection for disease resistance in herbage legumes at Svalof, Sweden. *Proc. Intern. Grassl. Congr.* 15:230–231. Kyoto, Japan.

Andersson, B. and V. Kristiansson. 1989. Breeding for resistance to root rot in red clover. *Proc. Intern. Grassl. Congr.* 16:363–364.

Arseniuk, E. 1989. Effect of induced autotetraploidy on response to *Sclerotinia* clover rot in *Trifolium pratense* L. *Plant Breeding.* 103:310–318.

Blain, F., M. Bernstein, S. Khanizadeh and S.A. Sparace. 1991. Phytotoxicity and pathogenicity of *Fusarium roseum* to red clover. *Phytopathology.* 81:105–108.

Chapurin, V.F., A.N. Afonin and G.I. Moskaleva. 1988. Evaluation of the resistance of red clover to physiological rot. *Sbornik Nauchnykh Trudov po Prikladnoi Botanike, Genetike, i Selektsii.* 120:73–79.

Gustavsson, A. 1988. Downy mildew on cultivated clover species in Sweden. *Vaxtskyddsnotiser.* 52:73–77.

Holub, E.B. 1988. Pathogenecity of *Aphanomyces euteiches* on *Medicago sativa* and evidence for a disease complex with *Phytophtora megasperma* f. sp. *medicaginis*. Ph.D. diss. Univ. of Wisconsin-Madison (Diss. Abstr. AWB-H7596-E753).

Jin, X., J.B. Kotcon and J.B. Morton. 1991. Interactions between *Pratylenchus penetrans* and *Fusarium avenaceum* in red clover. *Nematropica.* 21:105–109.

Kovacikova, E., A. Fojtik, H. Jakesova and A. Vlasakova. 1991. Testing red clover for resistance to the causal agents of root rots. *Sbornik UVTIZ, Ochrona Rostlin.* 27:63-72.

Kovacikova, E., H. Jakesova and F. Novotny. 1989. Amino acid contents in red clover varieties with varying levels to fungal pathogens of the genus *Fusarium*. *Rostlinna Vyroba.* 35:1169–1174.

Kovacikova, E. and V. Jarosik. 1993. Red clover response to *Fusarium oxysporum* and *Fusarium solani*, causal agents of crown and root rots. *Zentralbl. Microbiol.* 147:405-408.

Kovacikova, E. and V. Kudela. 1989. Relationship between the intensity of infection of red clover seeds by *Fusarium* spp. and plant resistance. *Sbornik UVTIZ, Ochrana Rostlin.* 25:17–21.

Leath, K.T. 1985. General diseases. pp. 205–233. In N.L. Taylor (ed.), *Clover Science and Technology.* ASA-CSSA-SSSA. Mono. 25:205–233. Madison. WI.

Leath, K.T. 1989. Diseases and forage stand persistence in the United States. pp. 465–479. In Marten, G.C., A.G. Matches, R.F. Barnes, R.W. Brougham, R.J. Clements, and G.W. Sheath (eds.), *Persistence of Forage Legumes.* ASA-CSSA-SSSA. Madison, WI.

Leath, K.T., F.L. Lukezic, B.W. Pennypacker, W.A. Kendall, R.G. Levine and R.R. Hill, Jr. 1989. Interaction of *Fusarium avenaceum* and *Pseudomonas viridiflava* in root rot of red clover. *Phytopathology.* 79:436–440.

Marum, P., R.R. Smith and C.P. Grau. 1994. Development of procedures to identify red clover resistant to *Sclerotinia trifoliorum. Euphytica.* 77:257–261.

Nagovitsyna, A.V. and N.N. Gracheva. 1990. Resistance of red clover and lucerne to *Fusarium* root rot. *Selektsiya i Semenovodstvo Moskva.* 3:28–30.

Nedelnik, J. 1986. Susceptibility of some red clover cultivars to selected *Fusarium* spp. and the effect of fungicides on growth of the *F.* spp. *in vitro. Sbornik UVTIZ, Ochrana Rostlin.* 22:285–293.

Nedelnik, J. 1988. Methods of testing the resistance of vegetatively propagated red clover (*Trifolium pratense* L.) to *Fusarium* spp. *Rostlinna Vyroba.* 34:1197–1202.

Nedelnik, J. 1989. Comparison of the response of red clover plants to infection with the fungus *Fusarium solani* under laboratory and field conditions. *Sbornik Vedeckych Praci Oseva, Vyzkum, Slechteni a Semenarstvi Koncern Praha, Koncernovy Vyzkumny a Slechitelsky Ustav Picninarsky, Froubsko u Brna.* 11:197–203.

Nikiforova, E.V. and Y.N. Suslova. 1986. Breeding red clover for resistance to anthracnose. Selektsiya zemovykh kultur na ustoichivost k boleznyam i neblagopriyatnym faktoram sredy V Volgo-Vyatskom regione. 1986:30–34.

Paplauskiene, V. 1988. Rapid methods of diagnosing resistance of red clover to *Fusarium* pathogens. *Sbornik Nauchnykh Trudov, Litovskii Nauchno Issledovalel skii. Institut Zemledeliya.* 58:60–67.

Pederson, G.A. R.R. Hill, Jr. and K.T. Leath. 1980. Host pathogen variability for *Fusarium*-caused root rot in red clover. *Crop Sci.* 20:787–789.

Putsa, N.M. and O.Y. Mezentseva. 1989. Use of artificial infection regimes in breeding red clover for resistance to *Fusarium. Sbornik Nauchnykh Trudov, Vsesoyuznyi Nauchno Issledovatel Skii Institut Kormov.* 42:52–56.

Pfeffer, H. and P. Luth. 1990. Effect of a red clover monculture on the anti-phytopathogenic potential of soil with regard to *Sclerotinia trifoliorum* Erikss. *Nachrichtenblatt Pflanzenschutz.* 44:214–216.

Raynal, G. 1983. Production an laboratoire d'apothecies de *Sclerotinia trifoliorum* Erikss pour l'evaluation de la resistance de trefle violet a la sclerotiniose. *Agronomie.* 3:369–373.

Raynal, G. and J. Picard. 1985. Laboratory production of the sexual stage of *Sclerotinia trifoliorum*, the winter crown-rot agent of red clover. Consequences for improving control methods. *Proc. Intern. Grassl. Cong.* 15:782–783.

Rufelt, S. 1985. Selection for *Fusarium* root rot resistance in red clover. *Ann. Appl. Biol.* 107:527–534.

Sauer, R. 1984. On the importance of clover rot *Sclerotinia trifoliorum* Erikss. on red clover and possibilities for control with vinclozolin. *Mededelingen van de Faculteit Landbouwwetenschappen. Rijksuniversiteit Gent.* 49:277–282.

Scott, S.W. and D.R. Evans. 1984. Sclerotia of *Sclerotinia trifoliorum* in red clover seed. *Trans. Brit. Mycol. Soc.* 82:567–569.

Skipp, R.A., M.J. Christensen and Z.B. Nan. 1986. Invasion of red clover (*Trifolium pratense*) roots by soil-borne fungi. *N.Z. J. Agr. Res.* 29:305–313.

Smith, R.R. and A.E. Kretschmer, Jr. 1989. p 541–552. Breeding and genetics of legume persistence. In Marten, G.C., A.G. Matches, R.F. Barnes, R.W. Brougham, R.J. Clements and G.W. Sheath (eds.), *Persistence of Forage Legumes.* ASA-CSSA-SSSA. Madison, WI.

Smurygin, M.A. and I.N. Vinnik. 1991. Evaluation of red clover for winter hardiness and resistance to root rots. *Selektsiya i Semenovodstvo Moskva.* 2:15–16.

Stutz, J.C., K.T. Leath and F.L. Lukezic. 1984. Virulence differences of *Fusarium roseum* isolates in red clover roots not explained by phytotoxins activity. *Canad. J. Microbiol.* 30:1494–1499.

Taylor, N.L., R.R. Smith and J.A. Anderson. 1990. Selection in red clover for resistance to northern anthracnose. *Crop Sci.* 30:390–393.

Tofte, J.E., R.R. Smith and C.R. Grau. 1991. Selection for resistance to *Aphanomyces euteiches* in red clover. *Crop Sci.* 31:1141–1144.

Uzik, M. 1983. Selection for persistence and root rot resistance in red clover. *Polnohospodarstvo.* 29:201–211.

Vanco, B. 1987. The effect of cutting on the development of root rot of red clover. *Rostlinna Vyroba.* 33:303–310.

Wanson, P. and H. Maraite. 1984. Root rot, the cause of red clover decline in Belgium. *Mededelingen van de Faculteit Landbouwwetenschappen. Rijksuniversiteit Gent.* 49:207–215.

7. Virus Diseases

Numerous viruses have been reported as infecting red clover. Thirty-four different viruses have been listed and doubtless the list could be extended with further inoculations (Edwardson and Christie, 1986). However, only seven viruses are of widespread major importance: Bean yellow mosaic potyvirus (BYMV), peanut stunt cucumovirus (PSV), red clover vein mosaic carlovirus (RCVMV), pea streak carlovirus (PStrV), alfalfa mosaic ilarvirus (AMV), white clover mosaic potyvirus (WCMV), and clover yellow vein potyvirus (CYVV) (Barnett and Diachun, 1985). They are mostly nonpersistent aphid transmitted viruses (Table 7.1). BYMV was the most prevalent virus infecting red clover in southeastern United States (McLaughlin and Boykin, 1988) and probably is the most important worldwide (Smrtz et al., 1983). Strain 204–1 of BYMV probably is the most prevalent strain, at least in the United States. Recently a new strain RC was found in Kentucky but it is not nearly as important as 204–1 (Doupnik et al., 1981). PSV is probably the second most prevalent virus of red clover at least in Kentucky (Naidu et al., 1995). The physical aspects of clover viruses have been described in detail by Barnett and Diachun (1986) and will not be dealt with here.

1. Effect of viruses on production

Three types of studies have been conducted on red clover. They include controlled environment and greenhouse studies, spaced plant studies, and small plot studies. Controlled environment studies conducted by Smith and Maxwell (1971) showed BYMV infected plants had less chlorophyll, shorter shoots, increased nitrogen content, higher percent moisture in stems, a higher leaflet to stem-ratio, and less dry matter per plant than healthy plants. Another study of this type was that of Khadhair et al. (1984) who examined the effect of WCM on various processes relevant to symbiotic N_2 fixation in red clover. Virus infection reduced plant growth, nodulation, rhizobial population, nitrogenase activity, and leghaemoglobin concentration in the nodules. RCVMV and MCMV in mixed inoculations reduced total dry matter yield by 57% in a greenhouse study (Bowen and Plumb, 1979). These kinds of studies are

Table 7.1

Virus diseases infecting red clover naturally*.

Virus	Group	Persistence	Vector	Symptoms**
Bean Yellow Mosaic Virus	Potyvirus	Nonpersistent	Aphid	Mosaic, mottle
Clover Yellow Mosaic Virus	Potyvirus	Nonpersistent	Aphid	Mosaic
Red Clover Vein Mosaic Virus	Carlovirus	Nonpersistent	Aphid	Mosaic, streak
Pea Steak Virus	Carlovirus	Nonpersistent	Aphid	Mosaic
Peanut Stunt Virus	Cucumovirus	Nonpersistent	Aphid	Mosaic
Alfalfa Mosaic Virus	Bacilliform particle	Nonpersistent	Aphid	Mosaic
White Clover Mosaic Virus	Potexvirus	—	Contact	Mosaic

* Adapted from Table 8.1 Barnett and Diachun, 1985.
** Symptoms may vary among plants, in addition to mosaic, and include mottling, necrosis, stunting, or no symptoms.

essential to indicate the specific components of plant growth affected by virus diseases but results cannot be reliably extrapolated to predict actual yield losses under typical field conditions (Campbell, 1986).

In several species of *Trifolium* naturally infected with BYMV, CYV, WCMV, and red clover mosaic rhabdovirus (RCMV), symptoms appeared early and were associated with loss of vigor and death of plants (Alconero et al., 1986). In clover plants of a space-planted nursery those with BYMV-like symptoms increased in percentage from 9.5 in Aug. 1980 to 29.8 in Sept. 1981 (Taylor et al., 1986b) (Fig. 7.1). In contrast, in plants of broadcast plots, no symptoms of virus diseases were present, and ELISA tests for BYMV and PSV of several symptomless plants were negative. The difference in infection between the spaced-plant nursery and the broadcast plots may have been due to the fact that the plant-soil mosaic of clean cultivated spaced-plant nurseries is more attractive to aphid vectors than are closed canopies. Therefore, such data should not be considered representative of field data because spaced plants are more easily infected than solid stands. Plots may be more representative of field situations where clover is often grown in mixtures with grasses. On the other hand, high incidences of virus diseases in broadcast fields have been observed frequently, but no evidence exists as to their effect on stand and yield.

Growing red clover in association with grasses also may affect virus infection. Mixing Italian ryegrass with the red clover cultivar, Hungaropoly, reduced the incidence of RCNV from 9 to 1% and of WCMV from 53.5 to

Fig. 7. 1. Early stages of virus infection in a spaced plant of red clover.

41% (Lewis et al., 1985). On the other hand, red clover yields may be reduced more by WCMV in grass clover mixtures than in pure stands (Scott, 1982). Taking this treatment one step further, Taylor and Ghabrial, (unpublished data) compared the cultivar Kenstar in spaced and broadcast plots with and without grass. Over the life of the stand in a spaced-plant nursery, Kenstar with grass had 9.7% infection with PSV and BYMV, as compared to 20.8 percent infection with no grass. In the broadcast plots, grass versus no grass had no effect as very few plants were infected as indicated by symptoms and ELISA. In conclusion, broadcast plot conditions particularly when clover is sown with grass are closest to the actual field environment of a pasture or hay field. However, the timing and incidence of natural virus infection may determine yield and stand maintenance. It is possible that precise data on economic losses from virus diseases will become available only after isogenic lines with and without virus resistance are developed.

2. Control of virus diseases

The most practical means of controlling virus diseases of red clover is via breeding (Taylor and Ghabrial, 1986a). The reader is referred to topics covered in that paper which include: Sources of resistance including resistance to vectors; epidemiology; genetics of resistance, screening for resistance; methods of breeding; freeing elite clones from viruses; interspecific hybridization and genetic engineering. Only the more recent literature will be discussed here. Breeding for resistance has been underway for many years. For example, in the United States, Kenstar and Arlington red clover cultivars have resistance to BYMV. In Czechoslovakia, BYMV-resistant cultivars were found among 23 cultivars and new selections tested in the field and in a greenhouse (Pokorny, 1989). In the field, diploids showed lower values for death from infection and for mean intensity of infection. In Kentucky (USA), a red clover germplasm resistant to strain 204–1 of BYMV was released (Taylor et al., 1985). The line was developed by crossing and backcrossing a clone homozygous for the hypersensitive type of resistance to each of the 10 clones of the Kenstar cultivar. After 5 backcrosses and one intercross generation, the 10 separate populations were blended to form a synthetic strain. In Pennsylvania (USA), 15 plants resistant to BYMV were selected from a mixed cultivar population of plants from 2, 3, and 4 year old fields and a population of Kenstar red clover plants grown in a greenhouse (Sim et al., 1985). The selected plants survived repeated inoculations and were considered resistant to strain 204–1 of BYMV. In Hungary, phenotypic mass selection was used to artificially select for resistance to BYMV in a colchicine-induced tetraploid population (Halasz et al., 1985). The selected material showed field resistance while maintaining its level of green matter production.

Inheritance of resistance to strains of BYMV has been investigated intensively. Resistance to the strain Strugi R104 was determined by a single dominant gene without cytoplasmic effects (Broda and Fredorow, 1984). A rather detailed group of studies on strain 204–1 of BYMV has been summarized by Taylor et al. (1986b). Two types of reactions occur in the Kenland cultivar; some plants become mottled, whereas other plants become necrotic. The necrotic reaction is controlled by a single factor (N) that is dominant to the mottling (n) reaction. Two types of necrosis, local and systemic also occur. In the hypersensitive response, necrotic spots are localized on inoculated leaves. This hypersensitive reaction is controlled by a gene (H) that is dominant over necrosis (h). Another type of resistance that shows no symptoms is controlled by a single factor (R) dominant over susceptibility (r), and apparently is epistatic to the hypersensitive factor.

For WCMV, tolerance in red clover behaved as if it were controlled by polygenes (Martin, 1989). Also, for RCVMV, resistance to a single isolate was controlled by a dominant gene (Kahn et al., 1978).

In summary, resistance to single strains of virus in diploid red clover is usually controlled by dominant genes. Resistance may be of the hypersensitive type, resistance to mottling, or difficulty of infection. Breeding has been conducted by backcrossing or by phenotypic recurrent selection, and resistance to several strains of each virus will be required for the desired level in a cultivar. The effect of virus diseases on yield and persistence in red clover remains to be determined.

Another method of increasing resistance to virus diseases in red clover is by transferring resistance from perennial species. The perennial species of *Trifolium* are remarkably resistant to viruses (Alconero et al., 1986; Taylor and Ghabrial, unpublished). The species *T. medium* and its *subspecies T. sarosiense* have been crossed with red clover and if sterility barriers of the hybrids can be broken, great potential for obtaining virus resistance exists (See Chapter 2, Biosystematics and Interspecific Hybridization). Likewise, genetic engineering has considerable potential for transfer of virus resistance to red clover.

References

Alconero, R., B. Fiori and W. Sherring. 1986. Relationships of virus infections to field performance of six clover species. *Plant Dis.* 70:119–121.

Barnett, O.W. and S. Diachun. 1985. Virus diseases of clovers. In N.L. Taylor (ed.), *Clover Science and Technology.* ASA-CSSA-SSSA. Mono. 25:235–268.

Barnett, O.W. and S. Diachun. 1986. Virus diseases of clovers: Etiology and crop losses. In J.R. Edwardson and R.G. Christie (eds.), *Viruses Infecting Forage Legumes.* S127 Mono. 14: Vol 3:625–675, Florida Agric. Expt. Sta. Gainesville.

Bowen, R. and R.T. Plumb. 1979. The occurrence and effects of red clover necrotic virus in red clover (*Trifolium pratense*). *Ann. Appl. Biol.* 92:227–236.

Broda, Z. and Z. Fiedorow. 1984. Inheritance of red clover (*Trifolium pratense* L.) resistance to bean yellow mosaic virus (BYMV). *Genetica Polonica.* 25:277–281.

Campbell, C.L. 1986. Quantifying clover yield losses due to virus diseases. 1986. In J.R. Edwardson and R.G. Christie (eds.), *Viruses Infecting Forage Legumes.* S127 Mono. 14, Vol 3:735–742. Florida Agric. Expt. Sta., Gainesville.

Doupnik, B., Jr., S.A. Ghabrial and N.L. Taylor. 1981. A new strain of bean yellow mosaic virus. *Phytopathology.* 71:871 (Abstr.).

Edwardson, J.R. and R.G. Christie. 1986. Viruses reported to infect certain forage legumes. In J.R. Edwardson and R.G. Christie (eds.), *Viruses Infecting Forage Legumes.* S127 Mono. 14, Vol 3:B1-B6. Florida Agric. Expt. Sta., Gainesville.

Halasz, E., L. Beczner, T. Gyori and S. Poczos. 1985. Phenotypic mass selection for resistance to BYMV and powdery mildew in red clover. *Novenytermeles.* 34:263–269.

Khadhair, A.H., R.C. Sinha and J.F. Peterson. 1984. Effect of white clover mosaic virus infection on various processes relevant to symbiotic N_2 fixation in red clover. *Canad. J. Bot.* 62:38–43.

Khan, M.A., D.P. Maxwell and R.R. Smith. 1978. Inheritance of resistance to red clover vein mosaic virus in red clover. *Phytopathology.* 68:1084–1086.

Lewis, G.C., A.J. Heard, R.A. Gutteridge, R.T. Plumb and R.W. Gibson. 1985. The effects of mixing Italian ryegrass (*Lolium multiflorum*) with perennial ryegrass (*L. perenne*) or red clover on the incidence of viruses. *Ann. Appl. Biol.* 106:483–488.

Martin, P. 1989. Genetic studies on resistance to alfalfa mosaic virus (AMV) and tolerance to white clover mosaic virus (WCMV) in red clover (*Trifolium pratense* L.) *Forage Notes.* 1989:34,80.

McLaughlin, M.R. and D.L. Boykin. 1988. Virus diseases of seven species of forage legumes in the southeastern United States. *Plant Dis.* 72:539–542.

Naidu, R.A., C.-C Hu, R.E. Pennington and S.A. Ghabrial. 1995. Differentiation of eastern and western strains of peanut stunt cucumovirus based on satellite RNA support and nucleotide sequence homology. *Phytopathology.* 85:502–507.

Pokorny, R. 1989. Response of some varieties of red clover to infection with bean yellow mosaic virus. *Sbornik Vedeckych Praci. Oseva, Vyzkum, Slechteni a Semenarstvi, koncern Praha, Koncernovy Vyzkumny-a-Slechtitelsky Ustav Picninarsky, Troubsko u Brna.* 11:205–213.

Scott, S.W. 1982. The effects of white clover mosaic virus infections on the yield of red clover (*Trifolium pratense* L.) in mixtures and in pure stands. *J. Agric. Sci. Camb.* 98:455–460.

Sim, S.T., K.T. Leath and C.P. Romaine. 1985. Evaluation of red clover for resistance to bean yellow mosaic virus. *Plant Dis.* 69:694–696.

Smith, R.R. and D.P. Maxwell. 1971. Productivity and quality responses of red clover (*Trifolium pratense* L.) infected with bean yellow mosaic virus. *Crop Sci.* 11:272–274.

Smrtz, J., M. Musil and V. Vacek. 1983. Virus representation and susceptibility of red clover varieties to infection. *Sbornik Vedeckych Praci Vyzkumneho a Slechtitelskeho Ustavu Picninarskeho v Troubsku u Brna.* 8:201–206.

Taylor, N.L., S. Diachun and S.A. Ghabrial. 1985. Registration of red clover germplasm resistant to bean yellow mosaic virus. *Crop Sci.* 25:714–715.

Taylor, N.L. and S.A. Ghabrial. 1986a. Breeding forage legumes for resistance to viruses. In J.R. Edwardson and R.G. Christie (eds.), *Viruses Infecting Forage Legumes.* S127 Mono. 14, Vol 3:609–623. Florida Agric. Expt. Sta. Gainesville.

Taylor, N.L., S.A. Ghabrial, S. Diachun and P.L. Cornelius. 1986b. Inheritance and backcross breeding of the hypersensitive reaction to bean yellow mosaic virus in red clover. *Crop. Sci.* 26:68–74.

8. Nematode Diseases

The role that nematode diseases play in stand persistence and yield reduction of red clover is not well defined. Few, if any, studies have been conducted where the effects of nematodes have been conclusively separated from those of soil borne fungal diseases and other factors causing loss of stands and yields. Hague (1980) abstracted data published by the Society of Nematologists (USA) (Anonymous, 1971) and estimated that on 5.4 million ha of clover leys in the United States, various nematodes cause losses of up to 6% annually equalling a value of $33,000,000. General surveys suggest that several different nematodes are often found associated with red clover. These findings suggest that nematodes may play a significant role in reducing production and contributing to premature stand decline of red clover plantings in various parts of the world. Other than a short review by Leath et al. (1985), no comprehensive treatment of effects of nematodes on red clover is available; consequently a more complete review of the literature will be included here.

Although many different nematodes have been identified as pests of red clover, species belonging to four genera of nematodes seem to be the principal pathogens. These are the root-knot nematodes (*Meloidogyne* spp.), clover cyst nematode [*Heterodera trifolii* (Soffert)], lesion nematode [*Pratylenchus penetrans* (Cobb) Filipjev & Schuurmans-Stekhovem], and stem nematode [*Ditylenchus dipsaci* (Kuhn) Filipjev]. Nematode damage in a given geographic region is however, greatly influenced by soil type (texture), previous cropping history, mean temperature, and rainfall.

1. Root-knot nematodes

From a worldwide perspective of root-knot nematodes found on red clover, the northern root-knot nematode [*Meloidogyne hapla* (Chitwood)] is probably the most frequently identified root-knot species. However, much of the research related to root-knot nematode disease on red clover has been centered in the southeastern United States and has focused on other root-knot nematode species, particularly peanut root-knot nematode [*M. arenaria* (Neal) Chitwood], southern root-knot nematode [*M. incognita* (Kofoid and

White) Chitwood], and javanese root-knot nematode [*M. javanica* (Treub.) Chitwood]. In the early 1900's, root-knot nematode (species not identified) infections were found which were severely injuring the affected red clover plants (Bessey, 1911). Subsequent studies conducted in the 1950's and 60's indicated that root-knot nematodes were serious pests of red clover. Research in Mississippi showed that red clover was more severely damaged by *M. incognita* and *M. incognita* var *acrita* than alfalfa (Bain, 1958). During this same time frame, research was initiated in red and white clover to identify sources of resistance to root-knot nematodes (Bain, 1959). Early results suggested that progress could be made in selecting for resistance to *M. incognita* in red clover although red clover populations were generally more susceptible to root-knot nematodes than white clover. Selection and breeding for resistance appears to be the only strategy feasible for control of root-knot nematodes on red clover since no nematocides are registered for use in pastures.

Other research in South Carolina compared the response of red clover to root-knot nematodes with that of 25 other *Trifolium* species including white and crimson clover (McGlohon and Baxter, 1958) . These researchers stated that, "the inoculated plants of all 25 species (15–18 plants of each species) were severely galled". They also concluded that a program of screening large numbers of seedlings would be necessary to identify any possible sources of resistant individuals. The relative susceptibility of several varieties of red clover to *M. hapla*, *M. arenaria*, *M. javanica*, *M. incognita*, and *M. incognita* var *acrita* was studied by Allison (1956) who reported that all were very susceptible to these five nematodes. However, none of these programs reported further breeding or selection of red clover populations resistant to root-knot nematodes.

Other reports from the 1960's and 70's suggested that root-knot disease was a world wide problem. Reports from Zimbabwe (Rhodesia) where over 70 cultivars or strains of red clover were screened for response to root-knot nematodes found that all had "roots almost totally covered with swellings" induced by *M. javanica* (Martin, 1961) . Other workers in New Zealand screened 2 lines of *T. pratense* for reaction to *M. hapla*, and reported that they were tolerant host (plants allowing nematode invasion and reproduction without suffering appreciable damage) (Yeates et al., 1973). More recently researchers in Santa Catarina State, Brazil have reported *M. incognita* attacking the red clover cv. Estanzuela 116, causing leaf chlorosis and death of infected plants (Kalvelage and Brose, 1990). In Latvia, Erenfelde (1984) found that *M. hapla* built up as much as 20 fold in one year under red clover. In Pakistan Anjum et al. (1986) have reported *M. javanica* infections on red clover. Research in Canada has shown that carrots (*Daucus carota* L.) grown in rotation with red clover can experience serious problems from northern root-knot nematodes (Diamond et al., 1991).

Few controlled field studies have been conducted to evaluate the reduction in yield or persistence of red clover caused by root-knot nematode. In a

Fig. 8. 1. Greenhouse screening for resistance to root-knot nematode showing healthy resistant plants (L) and dying susceptible plants (R).

greenhouse study of red clover grown in one-half gallon glazed crocks of steamed soil and infected with either *M. hapla* or *M. incognita*, the dry matter yield was significantly reduced by both nematodes (Chapman, 1960) . All plants in crocks infected with *M. incognita* died between 100–150 days. Most plants of *M. hapla* died during this period also. Injury was produced more rapidly by *M. incognita* than by *M. hapla*.

Breeding for resistance to root-knot nematodes have been limited by the small number of plants of a population which could be screened in a reasonable time period. Hussey and Barker (1973) developed a method for easily collecting root-knot nematode eggs for use as inoculum. This technique was combined with use of specialized plant growth containers (Conetainers®) to develop a method for rapid screening of large numbers of clover plants for response to root-knot nematodes (Quesenberry et al., 1986). Using this method as many as 5–6,000 plants can be grown on a single greenhouse bench, exposed to a uniform initial nematode density and evaluated for their response to root-knot nematodes 8–9 weeks after infesting (Fig. 8.1). About 1–2 days would be required by two workers for nematode inoculum preparation and 2–3 days for plant response evaluation based only on root gall indices and number of egg masses observed. Plants free of or with low numbers of root gall or egg masses are selected for further screening.

Using a combination of natural selection in the field in early stages of a selection program and greenhouse selection in later stages, Quesenberry et al., (1989) demonstrated that progress could be made in selecting for tolerance to root-knot nematode in red clover. They compared root galling and nematode egg-mass production caused by *M. arenaria* (MA), *M. incognita* (MI), and *M. javanica* (MJ) on an original population, a population produced after

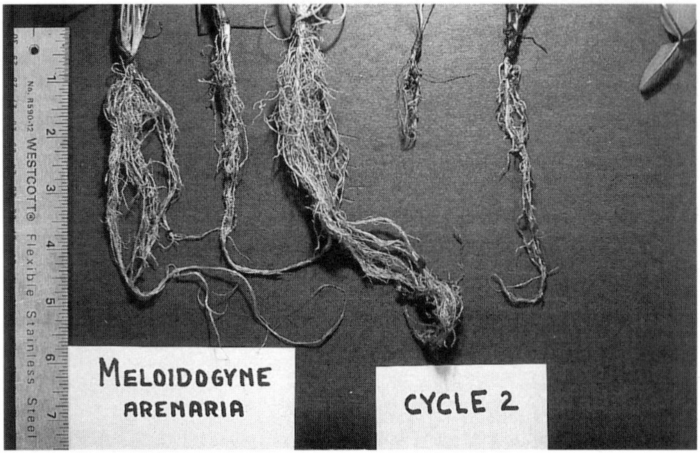

Fig. 8. 2. Variation in root galling response on red clover plants.

5 cycles of field selection, and a population produced after two additional cycles of greenhouse selection. Mean root gall indices (1–5 scale) of the base population and final population in response to MA, MI, and MJ were 4.5 vs 2.7, 4.4 vs 3.0, 4.2 vs 2.3, respectively. Egg mass scores, a measure of nematode reproduction, decreased more than galling. Mean egg mass scores (1–5 scale) of the base population compared to the final population for plants infested with MA, MI, and MJ, respectively, were 4.3 vs 2.2, 4.0 vs 1.8, and 4.0 vs 1.7. The intermediate population developed from this work (five cycles of field selection) was tested throughout the southeastern U.S. and released in 1992 as the cultivar 'Cherokee' (Quesenberry et al. 1993). Figure 8.2 shows the reduction in root galling between plants from an unselected root-knot susceptible population and plants in the root-knot resistant population. Further breeding and selection for root-knot nematode resistance has been conducted in this germplasm and a population showing high levels of resistance to MA, MI, and MJ is currently being tested. Preliminary results suggest that this population reduces numbers of juveniles maturing to adults and possibly reduces numbers of juveniles penetrating roots (Call and Quesenberry, 1994). A comparison of roots of a susceptible red clover and this breeding line showed fewer adults in roots of the resistant line at 8 weeks after infestation (Fig. 8.3).

Other research at Florida has reported screening about 400 plant introductions for response to *M. arenaria, M. incognita, M. hapla*, and *M. javanica* (Kouome et al., 1990). Results from this work showed that virtually all P.I.'s were susceptible to these root-knot species, although some variability for root galling and egg mass production was observed. These results further support the approach of recurrent selection within adapted cultivars to develop new cultivars with improved resistance to root-knot nematodes.

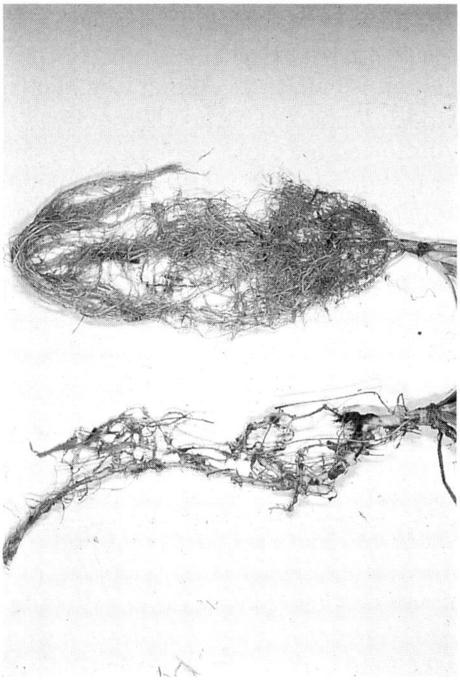

Fig. 8. 3. Reduction in numbers of juveniles maturing to adults in resistant red clover breeding line (top) compared to susceptible plants (bottom).

2. Root lesion nematode

Lesion nematodes [*Pratylenchus penetrans* (Cobb) Chitwood Oteifa] are often reported as parasites of red clover. Documentation of the reduction in yield or stand persistence caused by this nematode under field conditions is difficult to find. The effects of *P. penetrans* on the growth of 'Kenland' red clover growing in glazed crocks in a greenhouse in Kentucky has been reported (Chapman, 1958). In this study the plants were harvested 12 times, but after the 6th harvest the lowest level of nematode inoculum (500 per 1/2 gallon container) caused a significant reduction in yield. Number of surviving plants was reduced from 17 in the control to 8 in the low nematode inoculum treatment. This author further pointed out that the effect of lesion nematode was more severe during the hot summer months. In later research Chapman (1959) reported similar yield loss of red clover caused by *P. penetrans*.

The effects of lesion nematode on forage yield of red clover, white clover, alfalfa, and birdsfoot trefoil grown in a greenhouse at Prince Edward Island, Canada has also been reported (Willis and Thompson, 1969). This work found that red clover was second to birdsfoot trefoil in percentage yield reduction when inoculated with 750 nematodes per 4.5 l container. Recent survey research in Canada on carrots and crops grown in association has shown that *P. penetrans* was most prevalent on red clover and red clover timothy mixtures (Diamond et al., 1991). Other Canadian research has shown that root lesion nematode numbers in barley (*Hordeum vulgare* L.) roots were greatest when barley followed red clover-timothy. The combined dry weight of foliage and grain was reduced when barley was planted after barely or red clover-timothy (*Phleum pratense* L.) (Kimpinski et al., 1992). Red clover has also been shown to be a better host of lesion nematode than most associated grasses (Kimpinski et al., 1984).

Other researchers have shown that root-knot nematode may competitively inhibit *P. penetrans* invasion of red clover roots. Chapman and Turner (1975) and Turner (1971) suggest that previous invasion of roots by *M. incognita* may reduce the number and fecundity of *P. penetrans* on red clover. These findings may have little benefit on field production of red clover since *M. incognita* is generally thought to be equally or more damaging to clover yield and stand persistence than is *P. penetrans*.

Some researchers have suggested that in addition to the direct damage caused by *P. penetrans* to red clover, a more severe problem may be that the nematode creates wound sites for root fungal infections. Jin et al. (1991) evaluated the effects of *P. penetrans* and *Fusarium avenaceum* alone and in combination on growth of red clover. Although they did not find an interaction between these two pathogens, the greatest reduction in yield was at the highest level of inoculum of nematode and fungus combined. *P. penetrans* alone at an inoculum density of one nematode per cm^3 of soil significantly reduced red clover yield in this greenhouse experiment.

3. Clover cyst nematode

The clover cyst nematode (CCN) was reported in the United States on red clover in Illinois (Gerdemann and Linford, 1953). The extent of the distribution of this nematode and the severity of disease on red clover is not well documented. An early partial survey by Gerdemann and Linford (1953) found a *Heterodera* species which they identified as *H. schachtii* var *trifolii* (later changed to *H. trifolii*) [(Goffart) Oostenbrink] widely distributed in Illinois, but they reported only a few light infestations in fields of red clover. Later research by Mankau and Linford, (1960) proved the pathogenic effect of CCN on red clover, which was however, a less suitable host to CCN than white clover (*T. repens* L.). Chapman (1964) supports this finding and he states, "*H. trifolii* occurs in Kentucky widely distributed in low numbers associated with white clover, but is found rarely on red clover". A more recent survey from Prince Edward Island found CCN on red clover but at lower frequency than lesion nematode (Kimpinski et al., 1988). In Poland, 22% of red clover samples examined for plant parasitic nematodes contained cysts, but always at low population densities (Kornobis, 1987) .

Clover cyst nematode has also been reported parasitizing 'Grasslands Pawera' red clover in New Zealand within three weeks of overdrilling into mixed pasture containing white clover (Mercer and Campbell, 1986) . The clover cyst nematode was first reported in Pennsylvania in 1979 where premature stand decline of a red clover field was attributed to an infection of CCN on the roots (Leath et al., 1983) . After the initial stand was lost, a new seeding was made in cyst infested areas with and without a carbofuran, treatment. They reported that plants in non-carbofuran plots were severely stunted by cyst nematode activity, whereas with carbofuran establishment was comparable to that of non infected areas of the field. Forage yield in the year after seeding was significantly higher in the carbofuran treatment. They concluded that the CCN may be a significant factor in premature stand failure of red clover in other locations.

In greenhouse studies as few as 300 CCN second stage juveniles (J2) per quart (500 g) of soil caused significant reductions in yield of red clover compared to a non-inoculated control, and red clover was more severely injured by CCN than white clover. (Chapman, 1964). He suggested that the reason for the low degree of association of red clover and cyst nematode in field survey samples was that the nematode cannot maintain itself over a long period of time on red clover because the plants soon die. On the other hand, the nematode is often found in higher numbers on white clover but causes less severe damage, in terms of plant mortality. In Germany, glasshouse studies showed that 500 J2 per 100 cm^3 soil reduced yields by 15% and 5000 J2 per 100 cm^3 soil reduced yield by 50% (Stelter and Meini, 1972). Research from New Zealand where five different forage legumes including red clover were screened for response to CCN found that red clover was a tolerant host of *H. trifolii* (Yeates et al., 1973).

The variability among reports from different countries is likely related to differences in aggressiveness of the nematode populations, differences in susceptibility of red clover to the nematode, or different host races of CCN. Nematode populations with different aggressiveness to red clover have been reported (Singh and Norton, 1970). Marked variability in susceptibility to CCN has been observed among red clover cultivars and breeding lines (Norton and Isley, 1967). Field and microplot studies conducted by Maas and Heijbroek (1982) using the sugarbeet race of CCN (a newly identified host race in the Netherlands) indicated that this race reproduces poorly on red clover, although some reduction in crop yield was observed in infested microplots.

Research to select for genetic resistance to CCN has been conducted in Pennsylvania (Leath et al., 1985). The first cycle of selection in this research consisted of 15 plants from each of 5 adapted cultivars. These plants were surviving after two years in a field identified as heavily infested with CCN. The selected plants were intercrossed so that each plant served as a maternal parent in crosses with all other cultivars. Seed from these crosses produced 2000 progeny plants which were screened in the greenhouse for CCN response. After 0.5 years, 200 plants with no stunting, virus, or mildew symptoms were selected and polycrossed. A second cycle was produced in the same manner except that only 100 superior plants were selected. These 100 plants were intercrossed to produce the CCNR-1 germplasm (Leath et al. 1987).

CCNR-1 was compared, along with two cultivars not selected for resistance to CCN, for dry matter production of roots and herbage when grown with and without CCN feeding stress. After 16 weeks CCNR-1 yielded 58% of its unstressed control. The yield of CCNR-1 was significantly greater than yields of unselected cultivars that only yielded 31 and 42% of their controls. No other reports of breeding for resistance to clover cyst in red clover were found, although considerable work is in progress in New Zealand selecting for resistance to CCN in white clover (Van Den Bosch et al., 1993).

Other research has reported that 'Kenland' red clover and a number of other perennial and annual clovers were poor hosts of *H. glycines* Ichinohe (Windham and Lawrence, 1987). These workers suggest that these clovers can be used as cover crops on fields infested with soybean cyst nematode to reduce soil populations of this important pest. Work from Japan supports this finding (Shimizu et al. 1989). They reported that growing red clover for 2 or 3 years sharply reduced the *H. glycines* population on subsequent soybean crops. The mode of action for this observed reduction in *H. glycines* population may be that red clover is serving as a "trap crop". Roots of red clover are apparently readily penetrated by juveniles of *H. glycines* but maturation to adults does not occur (Schmitt and Riggs, 1991). Such effects would lead to reduction in soil populations of *H. glycines*. However, the variability in host rage aggressiveness which has been reported among CCN populations suggests that host preference of a population in a region should be evaluated before adopting red clover or any other species as a cover crop to reduce soil nematode populations.

4. Stem nematode

The most extensive damage by nematodes to pasture legumes may be caused by the stem nematode. At least 20 biological races of stem nematode occur and at least one race has red clover as its primary host (Hague, 1980). More recently Whitehead et al. (1987) attempted to characterize 67 stem nematode populations by reaction on a number of different plants. They found that alfalfa was resistant to all but alfalfa populations, and that multiplication of various populations varied greatly between cultivars of alfalfa and red clover. Some cultivars were resistant to some populations of alfalfa or red clover stem nematode and susceptible to others. These differences indicated the presence of different pathotypes or biotypes in different populations of the so-called red clover and alfalfa host races of stem nematode. Based on these results they recommend that new "resistant" cultivars should be tested against a range of populations before release for general use.

Various surveys of damage caused by stem nematode on red clover suggest that this pathogen is widely distributed. In the late 1940's and early 50's stem nematode was identified as a significant pest of red clover in all major farming areas of Sweden (Bingefors, 1956). Survey results from England conducted from 1953–55 suggested that stem nematodes were present in about 20% of clover fields and were causing moderate or severe stand loss in about 7% of the fields (Lester and Lange, 1958). These workers reported that natural selection in farmer fields in south Sweden resulted in the development of the resistant cultivar Merkur. Additional recent surveys from Sweden show that the frequency of stem nematode infested red clover fields varied from 19% (9 of 47) in Solleftea to 0% (0 of 67) in the Pitea area (Jonsson, 1981). An average of 6% (20 of 324 fields were infested. Infestation was more pronounced on heavier soil types. When four red clover cultivars differing in stem nematode resistance were infested with eight populations of stem nematode, no differences in pathogenicity were observed between the nematode populations and there was no evidence of resistance breaking. In Poland, 21% of red clover fields were infested with stem nematode (Glaba, 1990).

Other breeding and selection has resulted in the development of red clover lines with increased levels of stem nematode resistance (Bingefors, 1956). During the 1960's and 70's stem nematodes appear to have been identified as major problems on red clover throughout central and northern Europe. The optimum temperature for stem nematode appears to be 15–18 °C and these conditions are prevalent during the summer growing period in these areas. Programs of selection for stem nematode resistance in red clover have been active in Sweden (Jonsson, 1981), Denmark (Gus'kova, 1983); France (Caubel et al., 1972, 1985); Germany (Nuesch, 1989); Poland (Zakrzewski, 1977), and Russia (Putsa, 1982). Stem nematode has been shown to be transported on plant matter which may contaminate "home produced" red clover seed. Outside of Europe, Armstrong et al. (1977) state that stem nematode has been reported in many parts of the United States but that most damage occurs in the seed production areas of the western states (California, Idaho, Oregon, and Washington). Stem nematode has been associated with stand loss of 'Grasslands Hamua' red clover in New Zealand (Skipp and Christensen, 1990). These workers report that in one experiment most plants developed foliage symptoms of stem nematode during the first year. Stem nematode has also been reported on alfalfa, red clover, and sainfoin (*Oxobrychis viciifolia* scop.) in Iran (Sharafeh, 1987).

A modified technique for screening red clover seedlings for resistance to stem nematode has been reported (Toynbee-Clarke and Bond, 1970). This method was based on Dijkstra's (1956) modification of the method of Bingefors (1952). Bingefors has previously shown good correlation between susceptibility of artificially inoculated seedlings grown in greenhouse pots and field-trials on stem nematode infested land. Similar good correlation between field reduction in yield in stem nematode infected fields and greenhouse screening using the methods of Dijkstra have been reported (Fiddian and Aldrich, 1964). Using this method Toynbee-Clarke and Bond (1970) identified several single plant progenies and the tetraploid cultivar, Maris Leda, as being significantly more resistant (P = 0.01) than even the resistant control (Dorset Marl).

Other modifications of methods for screening for stem nematode resistance were used by Armstrong et al. (1977) in the United States to screen 682 selections of red clover. Lines which showed 40% or less infected seedlings in a preliminary screening were reevaluated in a second experiment and some lines with less than 10% infected seedlings were identified. The four most resistant entries came from Great Britain, Sweden, Denmark, and South Africa, however, none of the entries were immune. Nevertheless, this work further documented the range of variability for stem nematode resistance in red clover and provided additional germplasm sources which could be used for breeding and selection.

Various European researchers continue to focus on screening germplasm and developing new red clover cultivars with stem nematode resistance. Researchers in Poland reported screening 21 tetraploid cultivars for response to stem nematode using a greenhouse screening procedure (Zakrzewski and Malicka, 1978) . In addition to Maris Leda, they identified 'Ulva' and 'Pelly' as having high levels of stem nematode resistance. The diploid cultivar Weiball's Britta released by Swedish workers is reported to have high levels of resistance to stem nematode. Other resistant cultivars used in Scandinavia include Hermes, Resistant, and Merkur (Decker, 1981). Russian researchers have conducted field and laboratory studies of 20 local populations of red clover from Kaluga province, and reported identification of some populations with good resistance to stem nematode (Glushkov et al., 1981).

By the late 1970's most European cultivars of red cultivars had moderate to high levels of resistance to stem nematode. Bingefors (1985) reported the results of official Swedish test of 43 red clover cultivars evaluated between 1971 and 1979. In these test resistance to stem nematode was assessed in the laboratory on the basis of percentage of stunted seedlings. Only one of the 43 cultivars was rated as susceptible, and dry matter yield in the field was correlated with resistance to stem nematode. Breeding for resistance to stem nematode is continuing in Europe as evidenced by the release of 'Bjorn' (Wiberg, 1985), 'Sara' (Jonsson, 1985), and reports by Neusch (1989).

Although cultivars with high levels of resistance to stem nematode have been developed, populations of the stem nematode differ markedly in their virulence (Whitehead, 1992). He suggested that additional genes for resistance are needed and made a preliminary assessment of a wide range of *Trifolium* spp. He found that 13 species were very resistant (no symptoms of attacks, no *D. dipsaci* found in shoots) to a mixture of 4 strains of the 'red clover race' of stem nematode. However all these species were annuals and not sexually compatible with red clover. He did identify *T. diffusum* which can be hybridized with red clover as being resistant to stem nematode. Workers in France have also reported screening other *Trifolium* spp. for response for stem nematode and identified the annual species *T. resupinatum* Goss., *T. spumosum* L., and *T. squarrosum* L. as having resistance (Caubel et al., 1985).

References

Allison, J.L. 1956. Root-knot of perennial forage legumes. (Abstr.) *Phytopathology.* 46:6.

Anjum, M., A. Mohammed, B.A. Khan and M. Aslam. 1986. Some leguminous plants as new hosts of *Meloidogyne javanica. Pakistan J. Nemat.* 4:91–92.

Anonymous, 1971. Estimated crop losses due to plant-parasitic nematodes in the United States. Special Publication No. 1 Soc. Nemat. U.S.A. 7 pp.

Armstrong, J.M., J.N. Pinkerton and H.J. Jensen. 1977. Responses of red clover germplasm to stem nematodes in greenhouse trials. *Plant Dis. Reptr.* 61:1060–1063.

Bain, D.C. 1958. Reaction of red and white clover introductions to root-knot nematodes. (Abstr.) *Phytopathology.* 48:341.

Bain, D.C. 1959. Selection for resistance to root knot of white and red clover. *Plant Dis. Reptr.* 43:318–322.

Bessey, E.A. 1911. Root-knot and its control. U.S.D.A. *Bureau of Plant Industry.* Bul. No. 217:2–89.

Bingefors, S. 1952. Nematode resistance in clover and lucerne. *Proc. Intern. Grassl. Cong.* Pennsylvania, USA. 6:1591–1596.

Bingefors, S. 1956. Inheritance of resistance to stem nematodes in red clover. *Nematologica.* 1:102–108.

Bingefors, S. 1985. Improved resistance to stem nematodes in Swedish cultivars of lucerne and red clover and its value. *Zeitschrift fur Pflanzenzuchtung.* 95:164–172.

Call, N.M. and K.H. Quesenberry. 1994. Effect of three genotypes of *Trifolium* on the fecundity of *Meloidogyne arenaria* and *Meloidogyne javanica*. pp. 215–216. In *Agronomy Abstracts.* ASA-CSSA-SSSA, Madison, WI.

Caubel, G., F. Charpentier and P. Guy. 1985. Varietal resistance of forage legumes to *Ditylenchus dipsaci* (Kuhn) Fil. *Comptes Rendus des Seances l'Academie d'Agriculture France.* 71:719–728.

Caubel, G., D. Mugniery and R. Rivoal. 1972. Distribution de L'Anguillule des bulbes et des tiges,*Ditylenchus dipsaci* (Kühn) Fil., danslesol, a L'Intérieur d,un foyer d'attaque sur trèfle. *J. Zool. Écol. Anim.* 4:385–393.

Chapman, R.A. 1958. The effect of root-lesion nematode on the growth of red clover and alfalfa under greenhouse conditions. *Phytopathology.* 48:465–524.

Chapman, R.A. 1959. Development of *Pratylenchus penetrans* and *Tylenchorhynchus martini* on red clover and alfalfa. *Phytopathology.* 49:357–359.

Chapman, R.A. 1960. The effects of *Meloidogyne incognita* and *M. hapla* on the growth of Kenland red clover and Atlantic alfalfa. *Phytopathology.* 50:181–182.

Chapman, R.A. 1964. Effect of clover cyst nematode on growth of red and white clovers. *Phytopathology.* 54:417–418.

Chapman, R.A. and D.R. Turner. 1975. Effect of *Meloidogyne incognita* on reproduction of *Pratylenchus penetrans* in red clover and alfalfa. *J. Nemat.* 7:6–10.

Decker, H. 1981. Plant nematodes and their control (Phytonematology). USDA/NSF Joint Publication. Amerind Publishing Co. Pvt. Ltd. New Deli. 540 pp.

Diamond, J., J. Kimpinski, J. and C.E. Gallant. 1991. Root lesion and root-knot nematodes associated with crops grown in rotation with carrots on Prince Edward Island. *Canad. Plant Dis. Surv. Ottawa: Research Branch, Agric. Canada.* 71:13–15.

Dijkstra, J. 1956. Experiences with the breeding of red clover resistant to the stem eelworm. *Euphytica.* 5:298–307.

Erenfelde, E.-Ya. 1984. The northern gall nematode in the Latvian SSR. *Byulleten' Vesesoyuznogo Nauchno Issledovatel'skogo Instituta Zashchity Restenii.* 57:24–26.

Fiddian, W.E.H. and D.T.A. Aldrich. 1964. The susceptibility of red clover varieties to clover stem eelworm. *Plant Pathology.* 13:139–143.

Gerdemann, J.W. and M.B. Linford. 1953. A cyst-forming nematode attacking clovers in Illinois. *Phytopathology.* 43:603–607.

Glaba, B. 1990. Stem nematode on potato, red clover and beet in the Gdansk district. *Zeszyty Problemowe Postepow Nauk Rolniczych.* 391:53–63.

Glushkov, N.V., N.N. Gracheva, N.M. Putsa and G.F. Fateeva. 1981. Useful local populations of *Trifolium pratense*. *Selektsiya i Semenovodstvo USSR.* 1:24–25.

Gus'kova, L.A. 1983. Danish experience in testing varieties of clover and lucerne for resistance to *Ditylenchus dipsaci*. *Steblevye nematody sel'skokhozyaistvennykh kul'tur i mery bor'by s nimi (Materialy simpoziuma, Voronezh, 27–29 Sentyabrya 1983 g.)* pp. 136-137. Voronezh, USSR; Vserossiiskii NII Zashchity Rastenii.

Hague, N.G.M. 1980. Nematodes of Legume Crops. pp. 199–205. In R.J. Summerfield, R.J. Bunting and A.H.J. Richmond (eds.), *Advances in Legume Science* (Volume 1 of the Proc. of the International Legume Conference, Kew 31 July - 4 August, 1978).

Hussey, R.S. and K.R. Barker. 1973. A comparison of methods of collecting inocula of *Meloidogyne* spp. including a new technique. *Plant Dis. Reptr.* 57:1025–1028.

Jin, X., J.B. Kotcon and J.B. Morton. 1991. Interactions between *Pratylenchus penetrans* and *Fusarium avenaceum* in red clover. *Nematropica.* 21:105–110.

Jonsson, B. 1981. Occurrence of stem nematode (*Ditylenchus dipsaci*) in north Swedish ley farming, and pathogenicity tests of nematode populations. Examensarbeten. Uppsala, Sweden; *Sveriges Lantbruksuniversitet, Institutionen for vaxt och skogsskydd.* [viii], 41 pp.

Jonsson, H.A. 1985. Red clover (*Trifolium pratense*) Sara. *Agri. Hort. Genet.* 43:43–51.

Kalvelage, H. and E. Brose. 1990. Occurrence of *Meloidogyne incognita* (Kofoid and White, 1919) Chitwood, 1949. Parasitizing forage legumes at Santa Catarina State, Brazil. *Nematologia-Brasileira.* 14:141–142.

Kimpinski, J., L.M. Edwards, C.E. Gallant, H.W. Johnson, J.A. Macleod and J.B. Anderson. 1992. Influence of previous crops and nematicide treatments on root lesion nematode populations and crop yields. *Phytoprotection.* 73:3–11.

Kimpinski, J., H.T. Kunelius and B.N. Craid. 1988. Occurrence of plant-parasitic nematodes in forage legumes and grasses. *Forage Notes.* 32:30–32.

Kimpinski, J., H.T. Kunelius and C.B. Willlis. 1984. Plant parasitic nematodes in temperate forage grass and legume species in Prince Edward Island, Canada. *Canad. J. Plant Pathol.* 6:160–164.

Kornobis, S. 1987. Plant parasitic nematodes with poor growth of red clover in the region Wielkopolska. *Prace Naukowe Instytutu Ochrony Roslin.* 29:109–117.

Kouome, C.N., K.H. Quesenberry and R.A. Dunn. 1990. Response to root-knot nematode of a germplasm collection of red clover and related species. *Proc. Trifolium Conference.* Sublimity, OR. 9–11 June 1990.

Leath, K.T., J.R. Bloom, R.R. Hill, Jr., T.D. Kaufman and R.A. Byers. 1983. Clover cyst nematode on red clover in Pennsylvania. *Phytopathology.* 73:369.

Leath, K.T., J.R. Bloom and R.R. Hill, Jr. 1985. Selection for resistance to the clover cyst nematode in red clover. *Prog. Rep. Clovers Spec. Purpose Legumes Res.* 18:55. Madison WI, USA.

Leath, K.T., J.R. Bloom, R.R. Hill, Jr. and R.R. Smith. 1987. Registration of red clover germplasm line CCNR-1. *Crop Sci.* 27:1316.

Lester, E. and E.C. Lange. 1958. Surveys of clover rot with incidental observations on eelworm in clover: England and Wales 1953–55. *Plant Path.* 7:115–124.

Maas, P.W.T. and W. Heijbroek. 1982. Biology and pathogenicity of the yellow beet cyst nematode, a host race of *Heterodera trifolii* on sugar beet in the Netherlands. *Nematologica.* 28:77–93.

Mankau, R. and M.B. Linford. 1960. Host-parasite relationships of the clover cyst nematode, *Heterodera trifolii* Goffart. *Illinois Agric. Exp. Stn. Bul.* 667. pp. 33–36.

Martin, G.C. 1961. The susceptibility of clovers and trefoils to the common root-knot nematode *Meloidogyne javanica*. *Rhodesian J. of Agric.* 58:62–65.

McGlohon, N.E. and L.W. Baxter. 1958. The reaction of *Trifolium* species to the southern root-knot nematode, *Meloidogyne incognita* var. *acrita*. *Plant Dis. Reptr.* 42:1167-1168.

Mercer, C.F. and Campbell, B.D. 1986. Early invasion of overdrilled red clover seedlings by nematodes. *N. Z. J. Agric. Res.* 29:495–499.

Norton, D.C. and D. Issley. 1967. Cyst production of *Heterodera trifolii* on some leguminosae. *Plant Dis. Reptr.* 51:1017–1020.

Nuesch, B. 1989. Impact of selecting for resistance on yield and persistence of red clover and Italian ryegrass. *Landwirtschaft Schweiz.* 2:195–199.

Putsa, N.M. 1982. The clover stem nematode. *Zashchita Rastenii, Mowcow,* USSR. 5:52–53.

Quesenberry, K.H., D.D. Baltensperger and R.A. Dunn. 1986. Screening of *Trifolium* spp. for response to *Meloidogyne* spp. *Crop Sci.* 26:61–64.

Quesenberry, K.H., D.D. Baltensperger, R.A. Dunn, C.J. Wilcox and S.R. Hardy. 1989. Selection for tolerance to root-knot nematodes in red clover. *Crop Sci.* 29:62–65.

Quesenberry, K.H., G.M. Prine, O.C. Ruelke, L.S. Dunavin and P. Mislevy. 1993. Registration of 'Cherokee' Red Clover. *Crop Sci.* 33:208–209.

Schmitt, D.P. and R.D. Riggs. 1991. Influence of selected plant species on hatching of eggs and development of *Heterodera glycines*. *J. Nemat.* 23:1–6.

Sharafeh, M. 1987. Distribution and rate of infestation of alfalfa stem nematode *Ditylenchus dipsaci* (Kuhn, 1857) Filipjev 1936, in Fars, Kohkiluyeh and Boyrahmad provinces. *Entomologie et Phytopathologie Appliques.* 54:1–2.

Shimizu, K., S. Aiba and Y. Mitsui. 1989. Nematodes. B. Soybean cyst nematode of soybean. ASPAC Extension Bull. No. 302, pp. 7–9.

Singh, N.D. and D.C. Norton. 1970. Variability in host-parasite relationships of *Heterodera trifolii*. *Plant Dis. Reptr.* 60:1834–1837.

Skipp, R.A. and M.J. Christensen. 1990. Selection for persistence in red clover: influence of root disease and stem nematode. *N. Z. J. Agric. Res.* 33:319–333.

Stelter, H. and G. Meini. 1972. The effects of the infestation of red and white clover by *Heterodera trifolii* and *Heterodera galeopsidis*. *Archiv. fur Pflanzenschutz.* 8:463–470.

Toynbee-Clarke, G. and D.A. Bond. 1970. A laboratory technique for testing red clover seedlings for resistance to stem eelworm (*Ditylenchus dipsaci*). *Plant Path.* 19:173–176.

Turner, D.R. 1971. Infection of seedlings of alfalfa and red clover by concomitant populations of *Meloidogyne incognita* (Kofoid and White) and *Pratylenchus penetrans* (Cobb). Dissertation Abstr. International 32B (4).

Van Den Bosch, J., C.F. Mercer, J. Grant and I.K. Black. 1993. Clover cyst nematode resistance in white clover. Summaries of Invited Papers & Posters for XVII International Grassland Congress, Sessions 1–24. p. 70.

Whitehead, A.G. 1992. Sources of resistance to stem nematode, *Ditylenchus dipsaci* (Kühn) Filipjev, in species of *Medicago* and *Trifolium*. *Ann. Appl. Biol.* 120:73–81.

Whitehead, A.G., J.E. Fraser and A.J.F. Nichols. 1987. Variation in the development of stem nematodes, *Ditylenchus dipsaci*, in susceptible and resistant crop plants. *Ann. Appl. Biol.* 111:373–383.

Wiberg, A. 1985. Plant breeding at the Norrland branch of Svalof. *Sveriges Utsadesforenings Tidskrift.* 95:166–172.

Willis, C.B. and L.S. Thompson. 1969. Effect of the root-lesion nematode on yield of four forage legumes under greenhouse conditions. *Canad. J. Plant Sci.* 49:505–509.

Windham, G.L. and G.W. Lawrence. 1987. Host suitability of annual and perennial clovers to races of *Heterodera glycines. Prog. Rep. Clovers Spec. Purpose Legumes Res.* 20:46.

Yeates, G.W., W.B. Healy and J.P. Widdowson. 1973. Screening of legume varieties for resistance to the root nematodes *Heterodera trifolii* and *Meloidogyne hapla. N. Z. J. Ag. Res.* 16:81–86.

Zakrzewski, J. 1977. Observations on the occurrence, biology and pathogenicity of *Ditylenchus dipsaci* on red clover in Poland. Part III. Occurrence of *D. dipsaci* on red clover in Poland. *Hodowla Roslin, Aklimatizacja i Nasiennictwo.* 21:595–606.

Zakrzewski, J. and A. Malicka. 1978. The resistance of red clover varieties to the stem eelworm (*Ditylenchus dipsaci* Kunh). *Hodowla Roslin, Aklimatyzacja i Nasiennictwo.* 22:321–329.

9. Insect Pests

Numerous insects attack red clover including general feeders such as grasshoppers, crickets, armyworms and other caterpillars, and miscellaneous beetles. Fortunately, their attacks are sporadic, and other species of plants may be as attractive or more attractive to many of these insects than red clover. A considerable number of insects are primarily pests of red clover, or cause more damage to red clover than to other species. The biology and control of many of these pests were reviewed by Manglitz (1985) (Table 9.1). Only the more recent literature will be reviewed here.

1. Insects that feed on foliage and stems

The potato leafhopper (*Empoasca fabae* Harris), a major pest of alfalfa (*Medicago sativa* L.), also attacks red clover, particularly the non-pubescent forms (Fig. 9.1). Population dynamics of the species (Peterson et al., 1992) revealed that survival was not significantly different among host plants: red clover, alfalfa, and birdsfoot trefoil (*Lotus corniculatus* L.). However, developmental time was fastest on alfalfa, intermediate on trefoil and slowest on red clover, and at their peak, adults were more abundant on alfalfa than on the other two species. Simmons et al. (1984) also examined potato leafhopper development on different hosts, including alfalfa and red clover among others. The alfalfa hosts were among the most preferred for oviposition. Kenstar red clover caused a low ovipositional rate, low fecundity, and short adult survival. The non-pubescent cultivar Kuhn was less resistant than the pubescent Kenstar. In other studies, the glabrous cultivar Marino was less resistant to the leafhopper than pubescent cultivars Prosper I and Tristan (Choo et al., 1987). These data suggest that alfalfa is a preferred host of the leafhopper, and that resistance might be easier to obtain in red clover than in alfalfa due to slow development of the insect. The importance of pubescence in imparting resistance to the leafhopper in red clover remains to be elucidated.

The pea aphid *Acyrthociphon pisum* (Harris) is another sucking insect that is a primary pest of red clover (Fig. 9.2). In addition to the direct damage caused by feeding, the pea aphid also causes damage indirectly by transmitting

Table 9.1

Insects that primarily attack red clover, and their characteristics (Manglitz, 1985).

Insect common name	Scientific name	Authority	Feeding type	Attacks	Resistance	Other control
Clover leaf weevil	*Hypera punctata*	Fabricius	Chewing	Stems & leaves	Mammoth cultivars	Parasitic fungus
Yellow clover aphid	*Therioaphis trifolii*	Monell	Sucking	Stems & leaves	Germplasm N-2	Parasitic fungus
Pea aphid	*Acyrthociphon pisum*	Harris	Sucking	Stems & leaves	GP N-2	Parasitic fungus †
Potato leafhopper	*Empoasca fabae*	Harris	Sucking	Stems & leaves	American types	Pubescence
Meadow spittlebug	*Philaenus spumarius*	L.	Sucking	• Stems & leaves	None	Insecticides
Clover root curculio	*Sitona hispidulus*	Fabricius	Chewing	roots (larvae)	None	None
Clover root borer	*Hylastinus obscurus*	Marsham	Chewing	roots (larvae)	None	None
Lesser clover leaf weevil	*Hypera nigrirostris*	Fabricius	Chewing	leaves, heads & seeds, only on first crop	None	Removal of top growth
Clover head weevil	*Hypera meles*	Fabricius	Chewing	leaves, heads, & seeds	None	None
Clover seed midge	*Dasyneura leguminicola*	Lintner	Chewing	flowers	None	None
Clover seed chalcid	*Bruchophagas platypterus*	Walker	Chewing	seeds	None	Insect parasite

† The parasitoid *Aphidius ervi* recently has been identified from red clover in the former Czechoslovakia (Nemec and Stary, 1985).

Fig. 9. 1. Potato leaf hopper on pubescent stem.

Fig. 9. 2. Pea aphid on stem of alfalfa.

virus diseases, including bean yellow mosaic virus and peanut stunt virus (See Chapter 7, Virus Diseases).

Pea aphids may be controlled by insecticides and by parasites, and also by host plant resistance which may be the most economical method. A resistant

Fig. 9. 3. Weevils of red clover and other plants: R. to L. Clover leaf weevil, dock weevil, alfalfa weevil, lesser clover leaf weevil.

strain, "N-2" released by Gorz et al. (1979), lowered the net reproductive rate per generation and increased development time for the first instar as compared to two susceptible cultivars. However, mean generation time did not differ among the three red clovers (Zeng et al., 1993).

The mechanisms of resistance operating in N-2 were found to be a combination of antixenosis (nonpreference) and antibiosis. Tolerance apparently was not a factor inasmuch as no differences in plant growth were observed when equal numbers of aphids were maintained on the three clovers (Zeng et al., 1993). The authors concluded that such resistance would be useful for an integrated management program for pea aphids.

The pea aphid frequently is a pest of alfalfa. Experience on this crop apparently does not lead to host specialization, at least over a rather short period, and the aphid can easily shift to red clover (Via, 1991). Recently an unusual biotype was discovered in the former German Democratic Republic. The new biotype, which is red, still readily colonizes red clover, pea, and vetch (*Vicia villosa*) (Muller, 1984).

A relative of the pea aphid, the blue alfalfa aphid (*Acyrthociphon kindoi* Shinji) also attacks red clover in Nebraska (USA) according to a report by Rethwisch and Manglitz (1987). Apparently, the aphid does not reach epidemic proportions, perhaps due to predators. The predator *Adalia bipunctata* is reported to completely control unspecified aphids in some years in Russia (Gumovskaya, 1985).

2. Insects that feed on heads and flowers

The lesser clover leaf weevil (*Hypera nigrirostris* Fabr.) is an important pest of red clover seed fields. Its attacks are so severe that some populations do not produce blooms on the first growth in Kentucky (USA) (See Chapter 15, Seed Production). Fortunately, the biology of the insect is such that removal of the first growth for hay in May, and perhaps at other times, allows the second growth to develop free of the weevil. The weevil also feeds as larvae on foliage, and in the former German Democratic Republic has been reported to be the dominant species on red clover. Foliage losses over a period of several years ranged from 7.3 to 47.5% (Schumann, 1985).

In northeastern Saskatchewan (Canada) studies of the biology of the lesser clover leaf weevil showed that the adults overwinter in red clover fields, and begin to feed, mate, and oviposit in late April and early May. Eggs are laid in stipules, shoots, and leaves from early May to mid July. Larva are located on stipules, buds, and flowerheads. Pupa are present in late June until August, and new season adults first occur in July (Weiss and Gillott, 1993).

Another weevil that attacks red clover is the red clover weevil (*Phytonomus meles* Fabr.) also known as the clover head weevil (*Hypera meles* Fabr.). In Romania, the weevil was studied from 1979 to 1981 where it infested 25–30% of clover inflorescences. The weevil took 51–67 days to develop from egg to adult, and laid 126–238 eggs per female during a mean oviposition period of 79 days. The insect was controlled by insecticide treatments applied against *Apion* species (Brudea, 1986).

Stem weevils (*Apion* spp.) are occasionally destructive insects that damage red clover and other *Trifolium* species in eastern Europe, Poland, and Russia. The species *A. virens, A. apricans, A. flavipes*, and *A. seniculus* apparently are all involved but the greatest damage is produced by larvae of *A. virens* (Czerniakowski, 1986). In the spring, adults feed on leaves before ovipositioning in the 1st, 2nd, and 3rd stem internodes, and auxiliary shoots. Larvae that develop cause the greatest damage by mining in the stems.

The highest proportion of plants attacked by stem weevils was in the cultivar Start (79.5%) (Vanco, 1988). The economic threshold for damage to red clover varies from 4 to 10 adults per M^2 (Karavyonchi et al., 1986, Zhukovshii and Inovov, 1985). Another threshold for control action is 5 beetles per meter of row in crops with up to 3 green buds per plant (Schumann and Delinsky, 1989). There is one generation per year. Control measures include the use of insecticides and cutting the crop to a level of 6–8 cm at harvest (Zhukovshii and Inovov, 1985). Insecticides that have been effective in Poland include thiodan (endosulfan) (Vanco, 1988), owadofos (fenitrothion) and zolone (phosalone) (Czerneakowski, 1986). In Sweden, fenitrothion was generally effective although some differences existed in control among species (Bjorkman, 1987). In the former German Democratic Republic only preparations based on camphechlor are registered for seed weevil control

(Schumann and Delinsky, 1989). Other control measures include the use of parasitoids, *Treaspis floricola* and *Spinthuus dubius* which were recently found parasitizing *Apion apricans* larvae in Poland (Miczulski, 1988). Among 132 cultivars tested for resistance in Siberia, two were found to be completely resistant and four others showed only 3-4% infestation (Rustinov, 1988).

3. Root feeding insects

The clover root curculio (*Sitona hispidulus* F.) is thought to be involved in the root-rot complex of red clover (See Chapter 6, Fungal Diseases, and Chapter 10, Persistence). However, no definitive research on this interaction has been conducted. The insect is destructive in widespread areas as shown by research in Poland where the insect prefers white clover (*T. repens*) over red clover (Wiech and Wnuk, 1985). In Hungary, a related species, *S. humeralis* commonly feeds on alfalfa (lucerne) and oviposits almost as many eggs on red clover as alfalfa (578 vs 600 eggs per female). Threshold temperature for embryonic development was 10°C. Larval development lasted 15 to 25 days and the pupal stage 10 to 14 days. The first adults appeared in mid-July, entered diapause, and overwintered as adults (Nadasy and Saringer, 1987).

In West Virginia (USA), studies of 4 species of root feeding insects including *S. hispidulus* showed that during summer, movement was to adjacent woods. During spring and fall, greater numbers of adults moved toward red clover from adjacent woods and from red clover to adjacent pasture (Culik and Weaver, 1994). In Kentucky (USA), a laboratory nutrient solution that fostered nodulation of red clover and alfalfa supported higher populations of *Sitona* than did a solution which prevented nodulation. Likewise, in the field, use of nitrogen to suppress nodulation in alfalfa significantly lowered emergence of adults as compared to control plots (Wolfson, 1987). In New Zealand, *Sitona* has been shown to be a factor in seedling establishment. Ladino clover (T. *repens*) was preferred, followed by alfalfa and red clover (Barrett and Byers, 1992).

The clover root borer (*Hylastinus obscurus* Marsham) is another insect implicated in root decline of red clover. The volatile extracts of diseased red clover roots that were attractive to the root borer have been compared with previously identified foliar compounds (Kamm and Buttery, 1984). None of the attractive compounds were competitive with crude extracts of diseased roots when tested individually, but certain combinations were competitive with crude abstracts in laboratory tests. However, adults were not attracted to individual compounds or mixtures in preliminary field tests. The rationale of the investigators apparently was that roots were required to be diseased by fungi before becoming attractive to the borer. In contrast, the clover root borer has been implicated as the primary agent causing root rot decline in red clover, with the fungus *Fusarium avenaceum* being relegated to secondary importance (Jin et al., 1992). Fumigation, *Fusarium* inoculation and high phosphorus (P) levels were treatments compared to zero levels of the treatments. Root borers occurred naturally, and injury was greater at high P levels in both fumigation treatments, and was associated with greater mean tap root diameter in the second season following P treatment. The extent of mortality in red clover stands resulting from root borer injury in this two year study could not determined, but the number of second year plants with tap root injury and the magnitude of gallery development suggested that the root borer was the primary causal agent of root decline in the field plots. However, the authors failed to take into consideration that root decline occurs in areas where root borers are not present. Obviously, other factors, including stress, are involved in root decline and subsequent shortening of the life of red clover stands (See Chapter 10, Persistence).

References

Barratt, B.I.P. and R.A. Byers. 1992. Legume seeding feeding preferences of adult *Sitona hispidulus* (F.) (Coleoptara: Curculionidae). *Environ. Entomol.* 21:103–106.

Bjorkman, I. 1987. The susceptibility of species of Apion to pesticides. *Vaxtskyddsnotiser.* 51:118–122.

Brudea, V. 1986. Biological investigations on the red clover weevil (*Phytonomus meles* F.) *Analele Institutului de Cercetari pentru Cereale si Plante Technice Fundulea.* 53:387-391.

Choo, T.M., B.J. Connolly, J.E. Langille, R. Drapeau, B. Coulman, R.B. Walton, J.S. Bubar, B. Gogven, J. Madill, J.G. Proulx and D.T. Fairey. 1987. Marino red clover. *Forage Notes.* 31:64–65.

Culik, M.P. and J.E. Weaver. 1994. Seasonal crawling activity of adult root-feeding insect pests (Coleoptera: Curculionidae; Scolytidae) of red clover. *Environ. Entomol.* 23:68–75.

Czerniakowski, Z. 1986. Clover weevils (*Apion* spp.) occurring in plantations of red clover (*Trifolium pratense* L.) white clover (*T. repens* L.), and alsike clover (*T. hybridum* L.) in southeast Poland. *Roczniki Nauk Rolniczych, E. Ochrona Roslin.* 16:17–100.

Gorz, H.J. and G.R. Manglitz and F.A. Haskins. 1979. Selection for yellow clover aphid and pea aphid resistance in red clover. *Crop Sci.* 19:257–260.

Gumovskaya, G.N. 1985. The coccinellid fauna. *Zashchita Rastenni.* 11:43.

Jin, X., J. Morton and L. Butler. 1992. Interactions between *Fusarium avenaceum* and *Hylastinus obscurus* (Coleoptera: Scolytidae) and their influence on root decline in red clover. *J. Econ. Entomol.* 85:1340–1346.

Kamm, J.A. and R.G. Buttery. 1984. Root volatile components of red clover: identification and bioassay with the clover root borer (Coleoptera: Scolytidae). *Environ. Entomol.* 13:1427–1430.

Karavyanskii, N.S., V.P. Blinova and V.B. Zhezmer. 1986: Economic thresholds. *Zashchita Rastenii.* 8:30.

Manglitz, G.R. 1985. Insects and related pests. In N.L. Taylor (ed.), *Clover Science and Technology.* ASA-CSSA-SSSA. Mono. 25:269–294.

Miczulski, B. 1988. Hymenoptera found for the first time in Poland as parasitoids of larvae of the clover seed weevil, *Apion apricans* Herbst (Col. Curculionidoe). *Polskie Pismo Entomologiczne.* 58:685–687.

Muller, F.P. 1984. A red biotype of *Acyrthosiphon pisum* on pea. *Beitrage zur Entomologie.* 34:91–95.

Nadasy, M. and G. Saringer. 1987. Ecological studies on *Sitona humeralis* Steph (Col., Curculionidae) in Hungary. *Mededelingen van de Faculteit Landbouwwetenschappen Rijksuniversiteit Gent.* 52:171–177.

Nemec, V. and P. Stary. 1985. The genetic diversity of the parasitoid *Aphidus ervi* on the pea aphid. *Acyrothosiphon pisum* in Czechoslovakia. *Acta Entomologicia Bohemoslovacu.* 82:88–94.

Peterson, S.S., J.L. Wedberg and D.B. Hogg. 1992. Survival development and population dynamics of *Empoasca fabae* (Homoptera: Cicadellidae) on three legume hosts. *Great Lakes Entomol.* 25:9–14.

Rethwisch, M.D. and G.R. Manglitz. 1987. Distribution and parasitoids of the blue alfalfa aphid *Acyrthosiphon Kondoi* Shinji (*Homoptera aphidae*) in Nebraska. *Kansas Entomol. Soc.* 60:557–561.

Rustinov, V.I. 1988. Resistance of red clover varieties to *Apion apricans. Selektsiya i Semenovodstvo, Moscow* No.4:34.

Schumann, K. 1985. The importance of weevils as foliage pests of clover and lucerne. *Nachrichtenblatt für den Pflanzenschutz in der DDR.* 39:38–41.

Schumann, K. and H. Delinsky. 1989. On the incidence and monitoring of the shrew weevil (*Apion sp.*) in red clover seed crops. *Nachrichtenblatt für den Pflanzenschutz in der DDR.* 43:18–21.

Simmons, A.M., B.C. Pass and K.V. Yeargan. 1984. Influence of selected legumes on egg production, adult survival and ovipositional preference of the potato leafhopper. *J. Agric. Entomol.* 1:311–317.

Vanco, B. 1988. Harmfulness of the clover seed weevil, *Apion virens* (Coleoptera, Curculionidae) on red clover. *Sbornik Uvtiz, Ochrana Rostlin.* 24:119–126.

Via, S. 1991. Specialized host plant performance of pea aphid clones is not altered by experience. *Ecology.* 72:1420–1427.

Wiech, K. and A. Wnuk. 1985. Preference of the clover root weevil, *Sitona hispidulus* (F.) (Col. Curculionidae) in relation to various cultivars of white clover, *Trifolium repens* L. *Polskie Pismo Entomologiczne.* 55:187–194.

Weiss, R.M. and C. Gillott. 1993. The biology of the lesser clover leaf weevil, *Hypera nigrirostis* (Fabr.) (*Coleoptera: Curcilionidae*) on red clover *Trifolium pratense* L. in Saskatchewan. *Canad. Entom.* 5:831–837.

Wolfson, J.L. 1987. Impact of *Rhizobium* nodules on *Sitona hispidulus*, the clover root curculio. *Entomologia Experimentalis et Applicata.* 43:237–243.

Zeng, F., G. Pederson, M. Ellsbury and F. Davis. 1993. Demographic statistics for the pea aphid (*Homoptera: aphididae*) on resistant and susceptible red clovers. *J. Econ. Entomol.* 86:1852–1856.

Zhukovskii, S.G. and S.G. Ivanov. 1985. Stem weevils on clover. *Zashchita Rastenii.* 11:42.

10. Persistence

Red clover is botanically a perennial, although it often behaves as a biennial. The crop usually lives 3 to 4 seasons in most clover-growing regions of the world, and is generally conceded to be lacking in persistence. Persistence of red clover may be defined, for our purposes, as the ability to maintain a stand over a long period of time without depending upon reseeding or volunteering. The longevity, or life span of red clover has been the subject along with that of other legumes of considerable speculation (Smith and Kretschmer, 1988) primarily concerned with means of increasing or at least maintaining legumes in mixtures with grasses for hay, silage, or pasture. One of the major deficiencies of forage stands is the inability to maintain an adequate legume component. It is the purpose of this treatise to examine persistence in red clover, the factors that affect it, and finally, approaches to the development of greater persistence.

1. Development of persistence

It is axiomatic that to be persistent, a crop must be adapted to its environment. Red clover originated in Asia minor and southeastern Europe (Merkenschlager, 1934), and spread over most of western Europe and began to be cultivated about 1600. After 1650, the cultivated types were introduced into the remainder of Europe and into America and other countries. The environment where red cover is best adapted is a very temperate climate, with few extremes of heat or cold, and deep, fertile, well-drained soils of relatively high pH (Fergus and Hollowell, 1960). In this environment, red clover grows luxuriously with herbage dry matter yields up to 19 metric tons per ha per year. In well-drained soils, the tap root may reach 3.0 m at maturity. Stems reach 80 cm, and leaves are sparsely pubescent with short appressed trichomes (Nyffeler, 1976a,b). Stands have persisted as long as five seasons. Occasionally, stands of red clover are reported to last for 9 or more years, but invariably when seed from such a field is sown in cultivar trials, the clover persists only 3 to 4 years (Fergus and Hollowell, 1960).

When red clover is introduced into an area and grown extensively, a series of land races evolve that are adapted to these specific regions. In Sweden, for example, many land races were grown on the same farms for 50–100 years without admixture. The names of these land races have been preserved and the adapted genetic materials became the basis for organized plant breeding efforts (Bingefors and Akerberg, 1961).

In order to be persistent, the Swedish land races were resistant or at least tolerant to stem nematodes (*Ditylenchus dipsaci*) and to various diseases, as well as being drought and winter hardy. To be winter hardy, clover must be adapted to day-length, and is classified as early, medium-late and late. Early clover is adapted to shorter days, and late clover to longer days. Short-day adapted plants grown under long days will be forced into flowering at an early stage. Repeated cuts at flowering limit storage of root reserves and decrease winter hardiness and persistence (Anderson, 1986). This process is similar for all factors of the environment, and over time, adaptive forces determine the longevity (persistence) of red clover to be maximized at about 3 to 4 years.

When red clover was first introduced into North America, it was not particularly persistent. Early research was concerned with "clover sickness", and with "clover-sick" soils (Fergus and Hollowell, 1960). Gradually, the practices of liming soils to obtain the desired pH, and of applying fertilizer overcame the problem of clover-sick soils, and farmers began to save their seeds leading to the formation of many farmer ecotypes or landraces. By trial and error, it was learned that the climatic and edaphic factors that most suited red clover were located in the area now known as the red clover belt. This belt ranges from the southern Tennessee border north to Canada, to the western border of Missouri, Iowa and Minnesota and east to the Atlantic Ocean. South of this area, except for higher elevations, red clover is used as an annual and, on the Pacific Coast and adjacent states, is grown under irrigation.

At the turn of the 20th century, both double-cut (medium) and single cut (mammoth) types were grown, but gradually the single-cut, being less persistent and lower yielding, all but disappeared. The single-cut type, where it is adapted as in Canada, is very late and generally considered to be more persistent than the double-cut type. Another phenomenon was the disappearance of the original non-pubescent European types and emergence of the densely pubescent American types. At first it was thought that pubescence was the result of natural selection for resistance to the potato leafhopper (*Empoasca fabae* Harris) but it is now understood that, in addition, pubescence is a mechanism of stress tolerance. Thus, red clover became somewhat adapted to U.S. conditions particularly in the northeast and north central regions. USDA and experiment station plant breeders began to develop red clover cultivars based on farmer ecotypes, and to study means of increasing persistence in red clover for the United States. It was found that no one single cultivar was adapted to the whole region, so northern and southern red clover belt cultivars, exemplified by Lakeland and Kenland respectively, were developed and later other cultivars were developed for sub-regional areas.

2. Factors involved in persistence

2.1. *Fungal diseases*

In the United States, it soon became apparent that several major diseases were attacking red clover and affecting its persistence. In the South, southern anthracnose, caused by *Colletotrichum trifoliorum* B & E severely depleted stands. The cultivar Kenland was developed with resistance to the disease and now disease symptoms rarely appear, even if northern non-resistant cultivars are occasionally introduced south of their area of adaptation. Likewise, in northern U.S., resistance to northern anthracnose caused by *Kabatiella caulivora* (Kirchn.) Karak was developed in the cultivar Lakeland, and the disease was largely eliminated.

Resistance to single races of pathogens causing these diseases are governed by one or a few genes but complete resistance of a population requires several genes (Smith and Maxwell, 1973). Resistance to these diseases has maintained the persistence of red clover in the United States to about 3 to 4 seasons of production. However, it is doubtful that the potential persistence was increased; i.e., above the persistence of susceptible red clover in the absence of the diseases.

Other diseases that attack red clover include crown rot caused by *Sclerotinia trifoliorum* Erikss, black patch caused by *Rhizoctonia leguminicola* Gough and Elliott, rust caused by *Uromyces spp.*, powdery mildew caused by *Erysiphe polygoni* D.C. em Salm., and many others (Leath, 1985). Although any one of these diseases may cause catastrophic losses, epidemics are sporadic. Resistance to these diseases, while desirable, probably does not increase potential persistence, but does increase the reliability of stand maintenance.

The root rot complex, on the other hand, is thought by many investigators to be a main cause of lack of persistence in red clover. Pathogens including *Rhizoctonia,Fusarium, Pythium, Phytophora, Mycothecium, Colletotrichum, Cylindrocladium, Phoma, Gliocladium, Leptodiscus*, and *Codinea* (Leath 1985, 1988) are reported to be involved in the root rot complex. Red clover apparently is infected by one or more fungi early in life of the stand, rots develop and plants eventually succumb. Despite many attempts to breed for resistance, efforts have been thwarted by the multiplicity of and interactions of causal agents. Red clover, even in the apparent absence of fungal pathogens, exhibits “internal breakdown”, a progressive disintegration of the crown that ultimately leads to death of the plant (Cressman, 1967).

2.2. *Virus diseases*

Many different virus diseases have been isolated from red clover, and most of these are thought to seriously reduce stands and yields (Barnett and Diachun, 1985). The most important of these are bean yellow mosaic virus (BYMV), red clover vein mosaic virus (RCVMV), and peanut stunt virus (PSV) (Taylor and Smith, 1994). One or more of these viruses have been implicated in reducing persistence (Newton and Graham, 1960, Kahn et al., 1978, Taylor et al., 1986). Occasionally, viruses have been shown to be widespread in farmer fields (Jones and Diachun, 1976).

Considerable breeding for resistance to virus diseases in red clover has been undertaken. Resistant red clover plants have been discovered (Diachun and Henson, 1960; Stuteville and Hanson, 1964) and cultivars with some resistance to viruses have been released. However, isogenic lines with and without virus resistance have not been compared for persistence under farmer field conditions. Persistence of the present virus-resistant cultivars is no longer than 3 to 4 seasons.

2.3. *Insect pests*

Many different insect species attack red clover and some undoubtedly reduce persistence if populations are high (Manglitz, 1985). Two insects involved in the root-rot complex are reported to affect persistence of red clover. The clover root borer, *Hylastinus obscurus* (Marsham) and the clover root curculio, *Sitona hispidulus* (Fabricius) by their feeding on clover roots may directly reduce persistence, or may provide an avenue for infection by root rot organisms. However, as suggested by Leath and Byars (1973) and Leath (1988) the evidence is not at all clear for either insect, and the clover root borer likely plays only a minor role in the initiation of rot, because they usually infest a field after rot has already started. No one has shown that clover persistence is increased in the absence of these insects. In fact, in areas where the root borer is absent, clover still lives the usual 3 to 4 seasons.

2.4. *Nematodes*

Resistance to the stem nematode in Swedish red clover is a prerequisite for persistence. Some evidence exists that nematodes, particularly root-knot nematodes, may play a role in the nonproductivity and premature stand decline of red clover, particularly southern United States which is outside the area of greatest adaptability (see Chapter 8). On the other hand, similarly to insects, no evidence exists that freedom from nematodes alone would increase resistance beyond 3 to 4 seasons.

2.5. *Management*

Obviously, poor management including inadequate lime, fertilizer, and nodulation, among other factors, may lead to inadequate stand maintenance. An important factor in stand maintenance in northern climates is time of sowing. The plants must have the opportunity to develop strong rosettes before the onset of winter. In southern Sweden, red clover should not be sown later than July 15. Management of clover for seed vs hay production has been the subject of considerable investigation. Smith (1957) showed that plants that flowered and produced seed in the first season were susceptible to winter killing. Choo (1984) also found that non-flowering plants persisted longer than flowering plants. However, no evidence exists that lack of persistence of early flowering plants is a heritable character. Data from Kentucky indicates that if seed-producing plants are allowed to develop strong rosettes by removing top growth in the fall (by Sept. 15 in Kentucky), carbohydrate reserves are restored prior to the onset of low temperatures. Apparently, in the more northern locations, the growing season is too short to allow development of strong rosettes after seed production.

Seed production only on third-year stands has been justified on the assumption that the practice would result in selection for persistence. However, data from two different experiments on genetic shifts indicated no positive results from the practice (Taylor et al., 1979; 1991). On the other hand, the experiments indicated that decline in performance of seed lots increased outside the area of adaptation was associated with shifts in flowering response. The earlier types produced the most seed but were less persistent.

2.6. *Genetics of persistence*

The inheritance of persistence *per se* has been investigated infrequently in red clover. Usually investigators are concerned with presumed causal factors. Estimates of narrow sense heritability of persistence on an individual plant basis are usually low, ranging from 14.2 (Cornelius et al., 1977) to 17.1% (Anderson et al., 1974). Similar estimates on hybrid means are higher ranging from 36.3 to 38.9% suggesting that phenotypic recurrent selection for persistence on an individual plant basis is likely to be ineffective.

Adventitious roots have long been thought to affect persistence by sustaining plants after the tap roots have rotted. Vigorous plants in the beginning of the fourth season (persistent) have been shown to possess more adventitious roots than less vigorous plants (Montpetit and Coulman, 1991a). Cause and effect relationships were not established, however. Another investigation showed that narrow sense heritability of adventitious rooting was low, only 0.30. The data indicated that progeny testing would be necessary to insure successful selection of the trait (Montpetit and Coulman, 1991b).

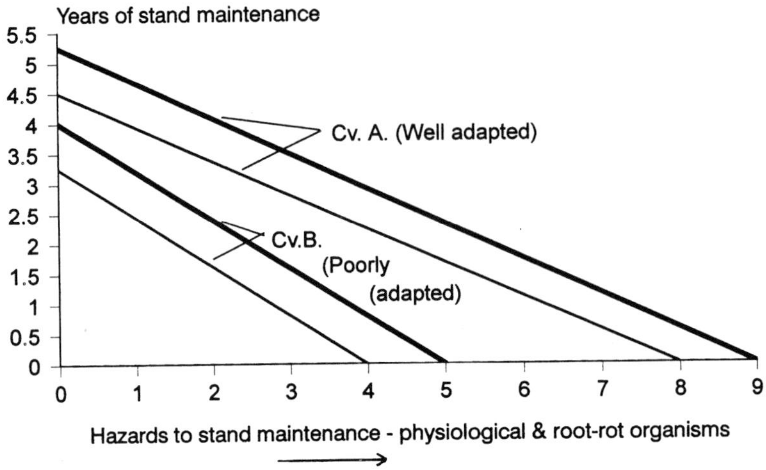

Fig. 10. 1. Relation of physiological hazards and root-rot organisms.

Another interesting development affecting persistence in red clover is the discovery of a stoloniferous type, that persists longer than non-stoloniferous types. However, heritability of this characteristic was not investigated (Smith and Bishop, 1993).

2.7. *Conclusions on persistence*

Although it is not a widely recognized concept, plants may be considered to have either closed or open reproductive systems (Molisch, 1938). In the open system, death of plant parts do not occur simultaneously due to various forms of vegetative reproduction. For example, live stolons of white clover perpetuate the plant even though the primary tap root is dead. These kinds of plants, also termed "guerilla" species, vary in the proportion of resources invested in stolons or rhizomes both seasonally and in different environments (Forde et al., 1988).

Red clover is a plant that is mostly of the closed system. As the plant ages, deterioration of the taproot begins even without pathogens (internal breakdown) and later saprophytic or weakly pathogenic organisms invade. Larger roots probably exhibit internal breakdown to a greater extent and earlier than smaller roots. Loss of the tap root results in death of the whole plant (unless adventitious roots are strongly developed, which is not usually the case in sown stands). Closed system plants generally have specified life cycles, controlled as in higher animals by unknown factors. Development of stolons (as reported by Smith and Bishop, 1993) would, of course, change the plant to an open system.

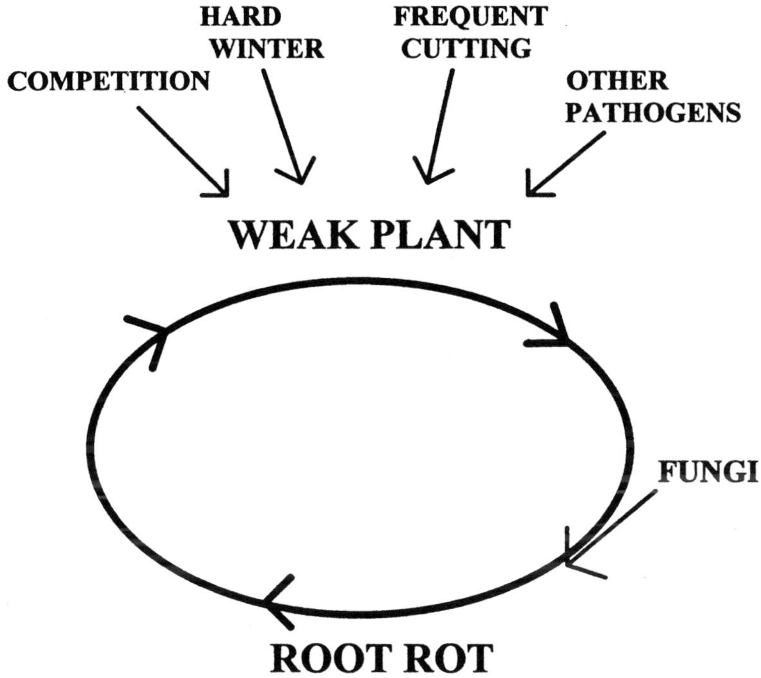

Fig. 10. 2. The vicious cycle of root-rot development (Modified from Rufelt, 1982).

The life cycle of red clover may, within limits, be increased by selection for greater adaptability which will delay senescence or increase persistence. The relationship of environmental factors and persistence (longevity) is illustrated in Fig. 10.1. (Taylor and Smith, 1977). As root-rotting organisms build-up in the soil or if the crop is exposed to poor management or other hazards, persistence will be shortened. Nevertheless, regardless of all other factors, senescence will eventually occur, and the stand will be lost.

Persistence, then, is the result of an interaction between the adaptation of the crop and its stress load. Stress load may be defined as any factor, whether physiogenic or pathogenic, that contributes to less than luxuriant growth. A well-adapted cultivar with little stress load would be expected to be strongly persistent for a time, but would eventually succumb due to senescence. On the other hand, a poorly adapted crop with the same or possibly even a lower stress load, would be weakened and the stand would die earlier. Support for this persistence (senescence) hypothesis comes from Sweden where it has been suggested that root rots in forage legumes may be an unavoidable disease (Rufelt, 1982). Environmental stresses; competition, winter hazards, management practices, and pathogens, are seen as reducing the vigor of the

Fig. 10. 3. Red clover plants selected for resistance or tolerance of root-rotting organisms: (R) Rotted tap root, (L) disease free tap root.

legume plants, (Fig. 10.2). Weakened plants are then susceptible to attack by other pathogens which increases root rot, which in turn further weakens the plant. Thus a "vicious circle" is set up with the result that root rot is unavoidable.

Plant breeders and others have tried to increase persistence by selecting for resistance to one or another of the root rotting agents (Fig. 10.3). Weakly pathogenic fungi such as *Fusarium* and *Leptodiscis*, the clover root borer (*Hylastinus obscurus* Marsham) and the clover root curculio (*Sitona hispidulus* Fabricius) fall into this category. Unfortunately, resistance to one of these pathogens may not increase persistence because other pathogens will still attack the crop. Many studies have shown that the crop may posses the bred pathogen resistance, but persistence still was not increased beyond the usual 3 to 4 seasons.

Malm and Hittle (1963), in a study of selection methods for improving persistence of red clover concluded that "screening for resistance to many root-rotting organisms naturally present in the red clover nursery was greatly superior to screening for resistance to the single pathogen, *Leptodiscus terrestris*." In breeding to increase persistence at a new location, it is necessary to breed for general overall adaptation, so that the crop will thrive and grow luxuriantly, and will be able to resist the many general pathogens and stresses for a long period of time. This may account for the initial success of many simple mass selection breeding programs begun at new locations. Occasionally, breeding under field conditions for such factors as increased forage yield or for grazing types, will inadvertently result in greater adaptation leading the breeder to believe that selection for a single character was successful.

Therefore, we may conclude that any program that increases adaptation in red clover should increase persistence within the limits imposed by the plant's life cycle. The most efficient way to do this is to expose crop plants to their environment and to select for survival. Breeding programs to increase persistence have been notably successful. Examples that may be cited include the Merkur cultivar in Sweden, the cultivars Hokuski and Tarsetsu in Japan and in the U.S., Arlington and Marathon in the northern red clover belt and Kenland and Kenstar in the southern part of the belt. The result is that red clover cultivars now will live usually up to 3 or 4 years in most countries, and occasionally longer if the stress load is not severe (Malm and Hittle, 1963; Taylor et al., 1962; Anderson et al., 1974).

To increase persistence beyond 3 or 4 years, greater concentration on breeding for general adaptation is necessary. Red clover has shown a considerable amount of hidden or latent variability. A classic case is the multiple head type. Although only double-parted or occasionally triple- and four-parted heads exist in red clover cultivars, six cycles of selection revealed heads with as many as 12 parts (Taylor et al., 1985). Persistence is a much less heritable character than multiple-parted heads for which annual phenotypic selection on individual plants was very successful. For persistence, selection on half-sib-family means, in microplots might be an effective procedure. Although persistence is usually expressed only in the third season, it might be possible to adjust the stress load to evaluate persistence after two seasons, at least initially. It is important to emphasize that longevity in red clover is controlled primarily by its life cycle, and that root rots or other agents are not the primary cause of lack of persistence.

Other methods that might be effective would be to change the plant to an open system of reproduction, as was previously mentioned. The development of stolons or rhizomes by introducing genes for these characteristics by conventional breeding methods, or by genetic transformation is a possibility.

Tetraploid red clover has been reported to be more persistent in some situations than diploids (See Chapter 13). Also, increase in persistence has been a long-time goal of interspecific hybridization programs (see Chapter 2).

In summary, persistence or longevity of red clover is primarily controlled by its natural life span as it is in many organisms with closed reproduction systems. Control of biological hazards and breeding effects are desirable to achieve the potential life cycle of red clover. However, increasing the longevity will necessitate genetic restructuring of the species.

References

Anderson, B. 1986. Effects of selection in herbage legumes. In Olsson, G. (ed.), Svolof 1886 - 1986. *Research and Results in Plant Breeding.* Svalof A B, Svalov, Sweden.

Anderson, M.K., N.L. Taylor and R.R. Hill. 1974. Combining ability in I_0 single crosses of red clover. *Crop Sci.* 14:417–419.

Barnett, O.W. and S. Diachun. 1985. Virus diseases of clovers. In N.L. Taylor (ed.), *Clover Science and Technology.* ASA-CSSA-SSSA. Mono. 25:235–268.

Bingefors, S. and E. Akerberg. 1961. Swedish land races of red clover. *Euphytica.* 10:147–151.

Choo, T.M. 1984. Association between growth habit and persistence in red clover, *Trifolium pratense. Euphytica.* 33:177–185.

Cornelius, P.L., N.L. Taylor and M.K. Anderson. 1977. Combining ability in I_1 single crosses of red clover. *Crop Sci.* 17:709–713.

Cressman, R.M. 1967. Internal breakdown and persistence of red clover. *Crop Sci.* 7:357–361.

Diachun, S. and L. Henson. 1960. Clones of red clover resistant to four isolates of bean yellow mosaic virus. *Phytopathology.* 50:323–324.

Fergus, E.N. and E.A. Hollowell. 1960. Red clover. *Advances in Agronomy.* 12:365–436.

Forde, M.B., M.J.M. Hay and J.L. Brock. 1988. Development and growth characteristics of temperate perennial legumes. pp. 91–109. In C.C. Marten, A.G. Matches, R.F. Barnes, R.W. Brougham, R.J. Clements and G.W. Sheath (eds.), *Persistence of Forage Legumes.* ASA- CSSA-SSSA Madison, WI.

Jones, R.T. and S. Diachun. 1976. Identification and prevalence of viruses in red clover in central Kentucky. *Plant Dis. Rep.* 60:690–694.

Kahn, M.A., D.P. Maxwell and R.R. Smith. 1978. Inheritance of resistance to red clover vein mosaic virus in red clover. *Phytopathology.* 68:1084–1086.

Leath, K.T. 1985. General diseases. In N.L. Taylor (ed.). *Clover Science and Technology.* ASA-CSSA-SSSA Mono. 25:205–233.

Leath, K.T. 1988. Diseases and forage stand persistence in the United States. pp. 465–479. In G.C. Marten, A.G. Matches, R.F. Barnes, R.W. Brougham, R.J. Clements, and G.W. Sheath (eds.), *Persistence of Forage Legumes.* ASA-CSSA-SSSA Madison, WI.

Leath, K.T. and R.A. Byers. 1973. Attractiveness of diseased red clover roots to the clover root borer. *Phytopath.* 63:428–431.

Malm, N.R. and C.N. Hittle. 1963. Selection methods for improving persistence in red clover. *Crop Sci.* 3:285–288.

Manglitz, G.R. 1985. Insects and related pests. In N.L. Taylor (ed.), *Clover Science and Technology.* ASA-CSSA-SSSA Mono. 25:269–294.

Merkenschlager, F. 1934. Migration and distribution of red clover in Europe. *Herb. Rev.* 2:88-92.

Molisch, H. 1938. *The longevity of plants.* Authorized English Version by E.H. Fulling. Publ. by the translator, New York.

Montpetit, J.M. and B.E. Coulman. 1991a. Relationship between spring vigor and the presence of adventitious roots in established stands of red clover (*Trifolium pratense* L.). *Canad. J. Plant Sci.* 71:749–754.

Montpetit, J.M. and B.E. Coulman. 1991b. Responses to divergent selection for adventitious root growth in red clover (*Trifolium pratense* L.) *Euphytica.* 58:119–127.

Newton, R.C. and J.H. Graham. 1960. Incidence of root-feeding weevils, root rot, internal breakdown and virus and their effect on longevity in red clover. *J. Econ. Entomol.* 53:865–867.

Nyffler, A., R. Koblet and J. Nosberger. 1976a. The effect of age on a stand of Swiss Mattenklee (*Trifolium pratense* L.) on dry matter production, leaf area, and stem growth. *Z. Acker-Pflanzenbau.* 142:214–225.

Nyffler, A., R. Koblet and J. Nosberger. 1976b. The effect of ploidy level on yield and morphological characteristics of Swiss Mattenklee (*Trifolium pratense* L.). *Z. Acker-Pflanzenbau.* 142:226–236.

Rufelt, S. 1982. Root rot - an unavoidable disease? A discussion of factors involved in the root rot of forage legumes. *Vaxtskyddsnotiser.* 6:123–127.

Smith, D. 1957. Flowering responses and winter survival in seedling stands of medium red clover. *Agron. J.* 49:126–129.

Smith, R.R. and A.E. Kretschmer, Jr. 1988. Breeding and genetics of legume persistence. pp. 541–552. In C.C. Marten, A.G. Matches, R.F. Barnes, R.W. Brougham, R.J. Clements and G.W. Sheath (eds.), *Persistence of Forage Legumes.* ASA-CSSA-SSSA. Madison, WI.

Smith, R.R. and D.P. Maxwell. 1973. Northern anthracnose resistance in red clover. *Crop Sci.* 13:271–273.

Smith, R.S. and D.J. Bishop. 1993. Astred-a stoloniferous red clover. *Proc. Intern. Grassl. Cong.* 17:421–423. Palmerston North, New Zealand.

Stuteville, D.L. and E.W. Hanson. 1964. Resistance to viruses in red clover. *Crop Sci.* 4:631–635.

Taylor, N.L., P.L. Cornelius and M.G. Long. 1985. Phenotypic recurrent selection for multiple-parted flower heads in red clover. *Crop Sci.* 25:489–494.

Taylor, N.L., S.A. Ghabrial, S. Diachun and P.L. Cornelius. 1986. Inheritance and backcross breeding of the hypersensitive reaction to bean yellow mosaic virus in red clover. *Crop Sci.* 26:68–74.

Taylor, N.L., R.G. May, A.M. Decker, C.M. Rincker and C.S. Garrison. 1979. Genetic stability of Kenland red clover during seed multiplication. *Crop Sci.* 19:429–434.

Taylor, N.L., C.M. Rincker, C.S. Garrison, R.R. Smith and P.L. Cornelius. 1991. Effect of seed multiplication regimes on genetic stability of Kenstar red clover. *J. Appl. Seed Prod.* 8:21 27.

Taylor, N.L. and R.R. Smith. 1977. Breeding for pest resistance in red clover, pp. 125–127. In *Proc. South. Past. and For. Crops Imp. Conf.* Sarasota, FL. June 13–14. ARS, USDA, New Orleans, LA.

Taylor, N.L. and R.R. Smith. 1994. Red clover. pp. 217–226. In R.F. Barnes, D.A. Miller, C.J. Nelson (eds.), Forages, Vol. 1. 5th edit. Iowa State Univ. Press. Ames, IA.

Taylor, N.L., W.H. Stroube, W.A. Kendall and E.N. Fergus. 1962. Variation and relation of clonal persistence and seed production in red clover. *Crop Sci.* 2:303–305.

11. Rhizobium Relationships

The literature on the microsymbiot for red clover now designated as *Rhizobium leguminosarum* biovar *trifolii* was reviewed by Burton (1985). His treatment pertained to all clover species whereas the present treatment will attempt to provide significant results since 1985 specifically for red clover.

1. Strain and host specificity

Some strains of the microsymbiot that are effective on red clover are also effective on strawberry clover (*T. fragiferum* L.), alsike clover (*T. hybridum* L.), ball clover (*T. nigrescens* Viv.), hop clover (*T. procumbens* L.), and white clover (*T. repens* L.). These species are termed subgroup B (Burton, 1985). Although some slight relationship exists between subgrouping and taxonomic classification (Zohary and Heller, 1984), the relationship is imprecise and microsymbiotic effectiveness is of little aid in taxonomic classification. Other species of the genus are divided into subgroups A and C through M but some species have not been assigned to a group (Burton, 1985). In fact, some species are not nodulated at all. These include *T. calcaricum*, *T. stoloniferum* and probably *T. virginicum* (Taylor et al., 1994 and personal observation). The groupings are not complete, that is, some strains of subgroup B may nodulate species outside the group. For example, some strains effective on red clover will also nodulate crimson clover (*T. incarnatum* L.) but usually this strain would not be as effective as a strain from subgroup A that includes *T. incarnatum* (Burton, 1985). However, strains of *Rhizobium* from Africa have been found to be effective on strawberry clover, a member of Burton's Group B, but ineffective on red clover, another member of the same group (Scheffler and Lovw, 1967).

Further strain specificity also occurs within the macrosymbiot. Red clover, being a cross pollinated species consists of many heterogeneous plants. The exact matching of *Rhizobium* strain with specific plants of clover suggests possibilities for genetic improvement of nitrogen fixation.

2. Inoculation

Most *Rhizobium* for inoculation of red clover is, like other inoculants, formulated and sold in a peat mixture. In the United States, the form usually supplied for convenience is that mixed with rhizobia effective on alfalfa (*Medicago sativa* L.), and the mixture will be effective on all *Trifolium* Group B species as well. Carriers other than peat such as vermiculite are possible and even preferred by some investigators (Paau et al., 1990). The use of commercial rhizobia has resulted in a build-up of strains in soil and the value of inoculation has begun to decline. Most red clover inoculant is now necessary only for new legume ground, or for insurance purposes (Paau et al., 1990).

Common methods of inoculation have been described as the sprinkle, slurry, or dry methods (Burton, 1985). Burton also described special methods of inoculation for hot dry soils, acid or alkaline saline soils, and soils heavily infested with native ineffective rhizobia. Seed pelleting is one such method, and research is still being conducted to determine its efficacy. For example, in Chile, red clover and other clovers were sown with various combinations of inoculation, pelleting, and applications of lime. In general, inoculation and pelleting increased nodule dry weight, nodule activity, dry matter production, and herbage nitrogen content but the effects were not always significant. In most cases, there was an interaction between inoculation and pelleting (Acuna et al., 1987). At the Kentucky station, three pelleted inoculants were tested on the Kenland red clover cultivar. No benefits were derived from the use of the commercially pelleted inoculum over the uninoculated control. Inasmuch as the treatments were applied to fields that had previously grown red clover, it is likely that the naturalized populations of rhizobia in the soil were more competitive than the introduced strains (Taylor, unpub.). Recently, commercial companies have marketed seeds with various film coatings and colorants to incorporate pesticides and seed protectants as well as *Rhizobium*.

3. Competitiveness of inoculant strains

A rhizobia strain with high competitiveness may be defined as one which has the ability to effectively nodulate red clover plants when more than one strain is present. Often, it is desirable that an introduced strain be competitive against indigenous strains. Genetic manipulations have enhanced rhizobia strains so that they have superior characteristics, and are highly efficient in nitrogen fixation in laboratory and greenhouse conditions. However, these new rhizobia proved not to be competitive in field situations due to their lack of competitiveness against indigenous strains (Paau et al., 1990).

This situation has led to considerable recent research to assess strain competitiveness and effectiveness. For example, in Canada, competition between indigenous and pure inoculant strains was examined under field conditions

(Ames-Gottfred and Christie, 1989). The indirect fluorescent-antibody technique was used to identify the nodule occupants. In the field, inoculations with 10 pure strains and a mixture of the strains were not able to completely overcome resident rhizobia which occupied, on the average, 50% of the nodules tested. Certain strains did result in increased dry weight and nitrogen content of red clover. No relationship was detected between strain effectiveness and competitive ability or between strain recovery and host cultivar. Concentration of indigenous strains was not considered to be a limiting factor in the recovery of introduced strains at this site (Ames-Gottfred and Christie, 1989).

In further research in Canada, repeated inoculation and increased inoculum levels increased dry matter production and red clover plant nitrogen content. Number of early nodules occupied by inoculant bacteria increased with increased inoculum concentration with one of two strains. However, occupying of late nodules was unaffected by inoculant levels in both strains (Martensson, 1990). In southern United States (Mississippi), dinitrogen fixation of 5 inoculant strains on red clover was examined over two seasons on two different soils. Three of the strains were highly competitive, persistent and accounted for greater than 90% of the assayed nodules on both soils over both seasons. Higher forage and crude protein production associated with the three strains led to the conclusion that competitive and persistent strains of the rhizobia can become established and productive in the lower south (Materon and Hagedorn, 1983).

In Sweden, under field conditions in a soil containing effective indigenous rhizobia strains, two inoculated strains were detected in minor proportions in nodules of established red clover using ELISA. However, no increase in dry matter production was obtained (Martensson et al., 1987). These examples of strain competitiveness suggest that strains can be isolated that will compete under some conditions with the indigenous field bacteria, and may ultimately result in greater dry matter yield and nitrogen content.

4. Amount of nitrogen fixed by *Rhizobia*

Although it has been long recognized that the symbiosis of *Rhizobium* and red clover produces large amounts of nitrogen, considerable research is still being conducted to determine amounts fixed in specific locations and circumstances (Table 11.1). Amounts of nitrogen fixed was determined by the acetylene reduction method, ^{15}N dilution, or by yields equivalent to that produced by applied N. The amount of N fixed ranged from 76 to 389 kg/ha in the various environments (Table 11.1). In most instances it was considered to be more economical to fix nitrogen than to apply it.

Table 11.1

Annual amount of nitrogen fixed, method of estimation, country and investigators conducting research on red clover.

Location	Annual amount fixed kg/ha	Method of estimation	Investigators
Czechoslovakia	187[a]	acetylene reduction	Simek, 1986
Germany	340	^{15}N dilution	Boller, 1988a
Germany	146[b]	^{15}N dilution	Boller, 1988b
Czechoslovakia	80–90	acetylene reduction	Machova, 1986
Canada	90–125	maize yield & equivalent	Bruulsema & Christie, 1987
New Zealand	76–105	acetylene reduction	Crush et al., 1983
Switzerland	308	^{15}N dilution	Boller and Nosberger, 1987
Uruguay	326–389[c]	^{15}N dilution	Mallarino et al., 1990

[a] Extrapolated from 210 days.
[b] Spring sown, lower when sown August.
[c] 1984–85 and 1983–84 respectively.

5. Effect of added N fertilizer on N_2 fixation

A considerable number of trials have been conducted to compare clover hay yields with *Rhizobium* plus applied nitrogen to those with *Rhizobium* alone. In plants grown in pots, hay yields were higher with extra N (Tabian and Moga, 1988) but evened out after low nitrate plants had increased their N-fixation. Also, in pot trials, extra levels of N increased yields but decreased nodulation (Posypanov et al., 1991). Results were similar with field trials. For example, in Germany, *Rhizobium* inoculation was compared with 0 and 40 kg N/ha as a starter fertilizer. Although inoculation and N application increased yields to a similar extent, for economic reasons, inoculation only was recommended for pasture containing in excess of 70% clover (Munsen and Schmude, 1989). Similarly, in Russia, application of 55 to 164 kg N/ha had no effect on yield in soils with pH 6.0 to 6.4 but decreased the proportion of red clover by 11 to 41% (Posypanov and Varokova, 1983). In Uruguay, results from field trials over a period of several years indicated that N applications increased herbage yield only in the first 18–20 weeks after application, and reduced fixed N yield (Mallarino et al., 1990). In New Zealand, the overall conclusion from a trilateral workshop on forage legumes for energy efficient animal production was that dry matter yields such as were attainable from good grass/*Trifolium* swards would only be obtained when pure grass swards were given 300 to 600 kg N/ha annually. It was suggested that N utilization following N_2–fixation may be more efficient than utilization from N fertilizer since leaching and run off from pasture treated with fertilizer may be considerable (Ball and Field,

1985). In conclusion, the results from most experiments throughout the world are that N applications to red clover may increase yield at first, but may delay N_2–fixation and reduce the proportion of clover in the sward. (See Chapter 4 also).

6. Effect of environmental conditions on N_2–fixation

Many environmental conditions adversely affect N_2–fixation in red clover. For example, in excess moisture (90% moisture capacity), nodules functioned less efficiently than at 60% moisture capacity (Fedorova 1988). Damage to bacteroids in nodules preceded death of the host cell.

Temperature also affects nitrogenase activity. In southern Finland, (60 °C north latitude), nitrogenase activity was still detectable at a soil temperature of 1.5 °C (air temperature 0.5 °C). Generally, nitrogenase activity and plant growth were correlated (Lindstrom, 1984). In Czechoslovakia, similar research showed that in the range of 10 to 21 °C, 1% change in temperature resulted in a 1% change in N_2-fixation activity as measured by ethylene production. Although this research establishes that nitrogenase activity is decreased by lower temperatures, strains of *Rhizobium* vary in this regard. Among strains isolated between 60 and 63°30′ N latitude in Finland, no strain grew *in vitro* at 5 °C but at other temperatures strains varied in growth (Lipsanen and Lindstrom, 1986) . This suggests that for optimum nitrogen fixation strain formulations of clover inoculant should be different for use in cool versus warm climates.

Red clover rhizobia are also affected by soil acidity. The lowest pH at which rhizobia grew on acid laboratory media was between 4.7 and 4.9 for 19 isolates from Finland (Lindstrom and Myllyniemci 1987). The pH of isolation sites of strains could be used to predict symbiotic performance in acidic soils. The best strain came from the most acidic isolation site. Conversely, nitrogen fixation by red clover, as well as other forage legumes is an important source of soil acidity in agricultural ecosystems. As soil acidity increases, nodulation decreases suggesting that applications of lime to raise pH levels to at least 6.3 is important (Posypanov and Varokova, 1985).

Other environmental effects include the application of fertilizers, especially molybdenum to soils with a deficiency of minor elements (Fedorova and Potatueva, 1984; Kudryashov, 1985; Collins and Lang, 1985, and Chalamet et al., 1987). Generally, addition of the element increased nitrogenase activity and hay yields.

7. Effect of pesticides on nitrogen fixation

The use of pesticides may directly affect nitrogen fixation through inhibitory effects on *Rhizobium* or indirectly by damaging the photosynthetic mechanisms of the red clover plant. In laboratory tests, the fungicides captan, fuberedoyole, HPMS and tilt severely damaged red clover. Other preparations that were not toxic included benomyl, quintozene, and thiophonate-methyl (Heinonen-Tanski et al., 1982). Herbicides and insecticides inhibited growth of rhizobia less than the fungicides. In other pure culture studies, hazardous agrochemicals and heavy metals at medium to high levels reduced nodule development (Martensson, 1992). Those reducing nodulations included benomyl, bentazone, chlorosulphoron, fenpropenorph, moncozeb, and monochlorophono-xyacetic acid. Dry matter production of nodulated plants was adversely affected by bentazone and chlorosulphoron. Chemicals that indirectly affected nitrogen fixation by having adverse effects on plant growth rather than on rhizobia included chlorosulphoron (Martensson and Nilsson, 1989) and dinoseb (Lindstron et al., 1985). In summary, it appears that many agrochemicals are hazardous to nitrogen fixation, particularly at higher levels of application. The exact amount to apply will have to be determined at a level that will have the desired function (weed, insect control, etc.) but still not severely injure the nitrogen fixation apparatus. Evaluations of new compounds should include measurement of effects on N_2-fixation.

8. Amount of N transferred to associated grasses

A few studies were conducted to not only determine the amount of N fixed but also to determine the amount or percentage transferred to associated grasses. In a study conducted in Uruguay, binary mixtures of red clover, white clover and birdsfoot trefoil with *Festuca arundenacea* were established at legume proportions of 30, 50, 70, and 90%. Nitrogen transfer from the legume to the grass was estimated by ^{15}N isotope dilution (Mallarino et al., 1990). Mean values for the percentage of grass N derived from the legumes were 20 for the first harvest, and later increased to 60. The annual uptake of transferred N averaged 18 and 34 kg/ha in the first and second years, respectively. It was concluded that these legumes transferred small amounts of N shortly after sowing, but transfer increased to significant values in the second year. Another study was carried out in Switzerland with red clover or white clover grown in association with rye grass (*Lolium multiflorum Lam.*) using the 15N dilution method. Apparent clover to grass transfer of fixed N contributed up to 52 kg/ha (17.5 N/ha on the average) to the N yield of the mixture (Boller and Nosberger, 1987). Percentage N derived from symbiosis averaged 75% for white and 86% for red clover. Further evidence of N transfer was provided by Wacquant et al. (1989) who studied $N0_3$-enrichment of a red clover - rye grass mixture (in pots) after $N0_3$ depletion. Their data suggested that $N0_3$ enrichment resulted from N excretion from active nodulated legume roots thus accounting for the increase in biomass and N content of the companion grass in a grass - legume association. In conclusion, it is apparent that symbiotically fixed N of red clover is transferred in significant amounts to companion grasses thus benefiting the association.

9. Breeding for increased N_2-fixation

Root-hair infection, nodule development and N_2 fixation of clovers are controlled by heredity factors in the host plant and in the rhizobia (Nutman 1984). Host genes that affect symbiosis in red clover include the factors *r* (host resistant to infection, no nodules), *d* (ineffective, no nitrogen fixed), *n* (ineffective, infection threads sparse) *ie* (ineffective, infection threads and nodule tumerized) *i*, (rapid nodule disintegration), *m* (normal nodule development), and *t* (normal nodule and bacteroid development). The progeny of crosses among *d,n ie*, and *i*, were all normally effective and the four factors segregated independently. The normally effective response was dominant over simply inherited ineffectiveness. Inbreeding increased the proportion of poorly effective plants, assumed by Nutman (1984) to be the result of the segregation of recessives. However, other explanations including loss of vigor are possible.

Very few breeding efforts to improve N_2–fixation of red clover have been undertaken. The complex nature of simultaneously selecting for efficient host plants and rhizobia is expensive and time consuming. An effort reported in the literature to improve the red clover host plant was that of Nutman (1984). In this program, plants of the cultivar S-123 were selected on the basis of yield after being inoculated with a highly effective strain, RCR5. Significant increases in yield under these conditions were reported. According to Nutman (1980), red clover bred for improved fixation has maintained its superiority against a range of *Rhizobium* genotypes. This suggests that breeding efforts for higher yield should, in effect, be breeding for high N_2–fixation provided that effective inoculants were involved. This agrees with results obtained in Switzerland in which genetic variation in fixed nitrogen was examined under three regimes of nitrogen fertilization: 0.5, 30, and 60N (kg/ha per cut). Although more nitrogen was fixed at lower levels of N fertilization, it was concluded that selection for increased dry matter was likely to result in enhanced yield of fixed N regardless of level of N fertilization (Boller and Nosberger, 1994).

In contrast to red clover, some success has been reported by Mytton (1984) working with alfalfa (*Medicago sativa* L.) and white clover (*Trifolium repens* L.), and by Barnes et al., (1984) working with alfalfa, all of whom emphasize the necessity of simultaneous improvement of many characters involved in the nitrogen fixation process including disease resistance, nodulation mass and foliar yield. Apparently, one of the most successful programs in breeding red clover for improved nitrogen-fixing ability has been achieved in Russia. Inoculation of light early maturing diploid forms resulted in increases of

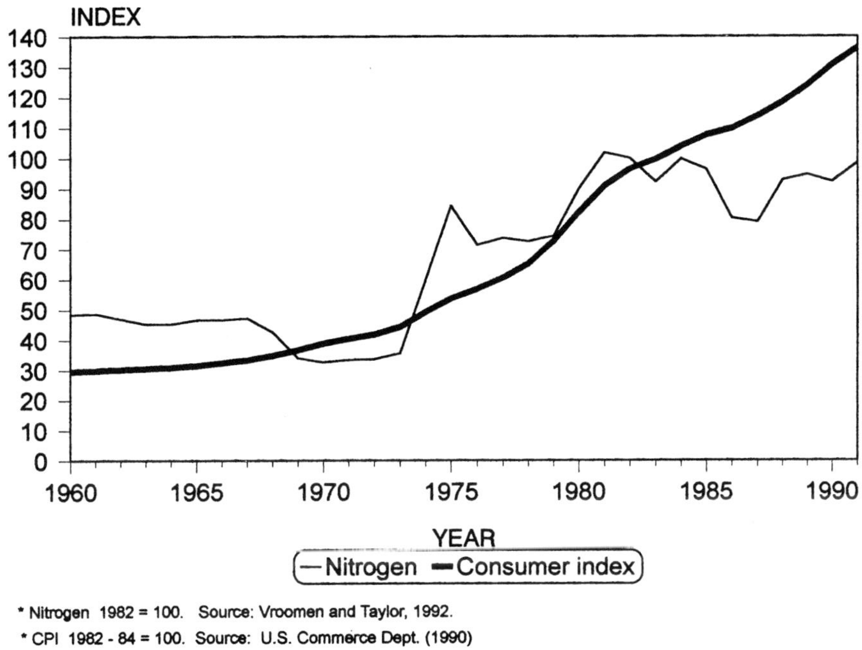

Fig. 11. 1. Nitrogen and consumer price indexes, 1960–91.

above-ground biomass of 103.5%, of dry matter yield of 187.6% and nitrogen fixed by 243.5% (Shain et al., 1988). The breeding scheme was described as being completed in only 6 years and was based on individual field evaluation of plots, selection, and microcloning of useful genotypes in vitro, inbreeding clones to stabilize traits, hybridizing to achieve heterosis, and field tests of hybrids (Kir'yan et al., 1991). Four promising early forms were bred which exceeded the uninoculated control over four years by 22.5 to 71.4% in dry matter yield, 8.3 to 56% in mean N_2-fixation/rate and 35 to 101.7% in crude protein yield.

10. Conclusions

Recent research on the red clover-rhizobia relationship suggests that considerable progress is being made. The matching of specific strains of rhizobia with specific cultivars of clover is an exciting possibility and can be done without modifying existing technology.

Another advancing area is that of competitiveness of inoculant strains. There appears to be ample evidence that new rhizobia strains can be developed that are both highly competitive against other inoculated strains and against indigenous field bacteria.

The amount of N that can be fixed by red clover rhizobia symbiosis is considerable, and most studies indicate that it is more economical to sow legumes to fix nitrogen than to apply it. The beneficial effects of the red clover symbiosis on quality of forage, tilth of soil, dry matter yields, as well as water quality are well documented (See Chapter 17). The question that might be asked, then, is why nitrogen continues to be applied rather than depending upon symbiotically fixed nitrogen. Primarily, the use of rotations in which red clover is a significant crop has declined. Secondly, red clover stands in pasture and hay fields are of short duration. To maintain dense stands of vigorous plants that have the highest N_2– fixation potential, it is necessary to renovate pastures every third or fourth season. This is a time consuming process, as well as expensive. A third factor is the availability of adequate supplies of nitrogen at low prices, at least in the United States. As shown in Fig. 11.1, since about 1983, nitrogen prices have not kept pace with the prices paid for most consumer goods (CPI) (Anonymous, 1990). This is probably due to stability of oil prices. As long as nitrogen prices are low it will continue to be cheaper and easier for farmers to apply N than to sow red clover. However, dwindling natural resources will eventually result in higher nitrogen prices, which together with environmental considerations may induce farmers to return to sowing legumes, including red clover. Other forage legumes may have more potential for grazing unless breeding and/or biotechnology efforts improve red clover and its microsymbiot. Very little research of this type is under way except perhaps in Russia. Based on their research, considerable potential exists for improvement of dry matter yield, and nitrogen fixation.

References

Acuna, P.H., O.A. Herrera and R.P. Gajardo. 1987. The effect of inoculation and seed pelleting on the establishment and nodulation of clovers. *Agricultura-Tecnica.* 47:51–56.

Ames-Gottfred, N.P. and B.R. Christie. 1989. Competition among strains of *Rhizobium leguminosarum* biovar *trifolii* and use of a diallel analysis in assessing competition. *Appl. Env. Microb.* 55:1599–1604.

Anonymous. Consumer Price Indexes. 1990. U.S. Commerce Department.

Ball, P.R. and T.R.O. Field. 1985. Productivity and economics of legume-based pastures and grass swards receiving fertilizer nitrogen in New Zealand. In Barnes, R.F. (ed.), *Forage Legumes for Energy Efficient Animal Production.* Proc. Trilateral Workshop, Palmerston North, N.Z. (1984).

Barnes, D.K., G.H. Heichel, C.P. Vance and W.R. Ellis. 1984. A multiple trait breeding program for improving symbiosis for N_2 fixation between *Medicago sativa L.* and *Rhizobium melioloti*. *Plant and Soil.* 82:303–314.

Boller, B. 1988a. Biological nitrogen fixation by white and red clover under field conditions. *Landwirtschaft Schweiz.* 1:251–253.

Boller, B., 1988b. Nitrogen fixation of Egyptian and Persian clover as compared with red clover. *Landwirtschaft Schweiz.* 1:309–312.

Boller, B.C. and J. Nosberger. 1987. Symbiotically fixed nitrogen from field-grown white and red clover mixed with ryegrasses at low levels of ^{15}N fertilization. *Plant and Soil.* 104:219–226.

Boller, B.C. and J. Nosberger. 1994. Differences in nitrogen fixation among field-grown red clover strains at different levels of ^{15}N fertilization. *Euphytica.* 78:167–174.

Bruulsema, T.W. and B.R. Christie. 1987. Nitrogen contribution to succeeding corn from alfalfa and red clover. *Agron. J.* 79:96–100.

Burton, J.C. 1985. Rhizobium relationships. In Taylor, N.L. (ed.), *Clover Science and Technology.* ASA-CSSA-SSSA Mono 25: 161–181. Madison, WI.

Chalamet, A., J.M. Audergon, J.P. Maitre and A.M. Domenach. 1987. Study of the effect of potassium on symbiotic nitrogen fixation by *Trifolium pratense* using the delta ^{15}N method. *Plant and Soil.* 98:347–352.

Collins, M. and D.J. Lang. 1985. Shoot removal and potassium fertilization effects on growth, nodulation and dinitrogen fixation of red clover (*Trifolium pratense*) and birdsfoot trefoil (*Lotus corniculatus*) under greenhouse conditions. *Field Crops Res.* 10:251–256.

Crush, J.R., G.P. Cosgrove and R.W. Brougham. 1983. Nitrogen fixation during 1979–1981 in two pastures on the Manawatu plains. *N.Z. Exp. Agric.* 11:17–20.

Federova, E.E. and Y.A. Potatueva. 1984. Ultra-structure and nitrogen fixation activity in red clover nodules on application of molybdenum. *Fiziologiya Rastenii.* 31:1120–1126.

Fedorova, E.E. 1988. Nitrogen-fixing system of red clover under waterlogging. *Izv. Akad. Nauk. SSSR Ser. Biol.* 0. 1987:272–277.

Heinonen-Tanski, H., G. Oros and M. Kecskes. 1982. The effect of soil pesticides on the growth of red clover rhizobia. *Acta Agriculturae Scantinavica.* 32:283–288.

Kiryan, I.G., A.A. Trukhan and A.S. Shain. 1991. Red clover as an object of non-traditional breeding for increased symbiotic nitrogen fixation. *Materialy knoferentsii molodykh uchenyhh i aspirantov "Novye idei v rastenievodstve i puti ikh realizatsii, Voronezh.* 9-13 iyulya 199:80–81.

Kudryashov, V.S. 1985. The effect of molybdenum on red clover yield. *Kormoproizvodstvo.* No. 2:37.

Lindstrom, K. 1984. Analysis of factors affecting in situ nitrogenase acetylene reduction activity of *Galega orientalis, Trifolium pratense*, and *Medicago sativa* in temperate conditions. *Plant and Soil.* 79:329–342.

Lindstrom, K. and H. Myllyniemi. 1987. Sensitivity of red clover rhizobia to soil acidity factors in pure culture and in symbiosis. *Plant and Soil.* 98:353–362.

Lindstrom, K., J. Polkunen and P. Kansanen. 1985. Effect of dinoseb on nitrogen fixation of red clover (*Trifolium pratense*). *Soil Biology and Biochem.* 17:865–869.

Lipsanen, P. and K. Lindstrom. 1986. Adaption of red clover rhizobia to low temperatures. *Plant and Soil.* 92:55–62.

Machova, M. 1986. Seasonal pattern of biological fixation of atmospheric nitrogen in red clover during the first and second cropping year. *Rostlinna Vyroba.* 34:907–917.

Mallarino, A.P., W.F. Wedin, C.H. Perdomo, R.S. Goyenola and C.P. West. 1990. Nitrogen transfer from white clover, red clover, and birdsfoot trefoil associated grass. *Agron. J.* 82:790–795.

Martensson, A.M. 1990. Competitiveness of inoculant strains of *Rhizobium leguminosorum* bv. *trifolii* in red clover using repeated inoculation and increased inoculum levels. *Canad. J. Microb.* 36:136–139.

Martensson, A.M. 1992. Effects of agrochemicals and heavy metals on fast-growing rhizobia and their symbiosis with small seeded legumes. *Soil Biol. and Biochem.* 24:435–445.

Martensson, A.M., J.G. Gustafsson and H.D. Ljunggren. 1987. Competition in soil between effective and ineffective strains of *Rhizobium leguminosarum* biovar *trifolii* in the nodulation of red clover, studied with ELISA. *J. of Gen. Microb.* 133:3467–3472.

Martensson, A.M. and A.K. Nilsson. 1989. Effects of chlorsulfuron on *Rhizobium* grown in pure culture and in symbiosis with alfalfa (*Medicago sativa*) and red clover (*Trifolium pratense*). *Weed Sci.* 37:445–450.

Materon, L.A. and C. Hagedorn. 1983. Competiveness and symbiotic effectiveness of 5 strains of *Rhizobium trifolii* on red clover (*Trifolium pratense*). *Soil Sci. Soc. Am. J.* 47:491–495.

Munsen, C. and D. Schmude. 1989. The yield efficiency of seed inoculation with *Rhizobium* in red clover/grass mixtures. Wissenschaftliche Zeitschrift der Wilhelm Pieck Universität Rostock. *Naturwissenschaftliche Reche.* 38:11–15.

Mytton, L.R. 1984. Developing a breeding strategy to exploit quantitative variation in symbiotic nitrogen fixation. *Plant and Soil.* 82:329–335.

Nutman, P.S. 1984. Improving nitrogen fixation in legumes by plant breeding, the relevance of host selection experiments in red clover (*Trifolium pratense*) and subterranean clover (*Trifolium subterraneum*). Proceedings of an FAO/IAEA consultant's meeting on breeding legumes for enhanced symbiotic nitrogen fixation. *Plant and Soil.* 82:285-301.

Nutman, P.S. 1980. Adaptation in nitrogen fixation. In W.D.P. Stewart and J.R. Gallon (eds.), *Proc. Phytochem. Soc. Europe Symp., Sussex.* pp. 335–354. Academic Press, London.

Paau, A.S., M.L. Bennett, C.J. Kurtenbach and L.L. Graham. 1990. Improvement of inoculant efficiency by strain improvement and formulation manipulations. In Greshoff, Roth, Stacey, and Newton (eds.), *Nitrogen Fixation: Achievements and Objectives.* Chapman and Hall. New York. London.

Posypanov, G.S., T.V. Batkova and V.I. Chernova. 1991. Biological fixation of nitrogen from the air by red clover as a function of composition of phytocenosis and the level of mineral nutrition. *Investiya Timiryazevskoi. Sel'skokhozyaistvennoi Akademii.* 1991:59–65.

Posypanov, G.S. and T.V. Voronkova. 1985. Effect of soil acidity and nitrogen fertilizers on red clover symbiosis with nodule bacteria and its yield in mixed stands with timothy. *Izvestiya Timiryazevskoi Sel'skokhozyaistvennoi Akademii.* 1985:186–189.

Posypanov, G.S. and T.V. Varonkova. 1983. Yield formation, symbiotic and photosynthetic activity of red clover in the second year with application of nitrogen fertilizers and irrigation. *Investiya Timiryazevskoi. Sel'skokhozyaistvennoi Akademii.* 1983. No. 2. 29–36.

Scheffler, J.G. and H.A. Lovw. 1967. The symbiotic characteristics of red clover rhizobia in the soils of the Stellenbasch district. *S. Afr. J. Agric. Sci.* 10:395–402.

Shain, A.S. and V.S. Majashenko. 1988. Breeding red clover for improved biological yield and nitrogen fixing ability. *Doklady Vsesoyuznoi Ordena Lenina i Ordena Trudovogo Krasnogo ZNameni Akatemii Sel'skokhozyaistvennykh Na uk Imeni V.I. Lenina* 9:31-33.

Simek, M. 1986. Nitrogenase activity (symbiotic nitrogen fixation) in red clover (*Trifolium pratense*). *Rostlinna Vyroba.* 34:919–926.

Tabian, I. and I. Moga. 1988. Influence of symbiotic and combined nitrogen on lucerne and red clover growth. *Analele Institului de Cercetari Pentru Cereale si Plante. Tehrice, Fundulea.* 56:391–400.

Taylor, N.L., J.M. Gillett, J.N.N. Campbell and S. Berger. 1994. Crossing and morphological relationships among native clovers of eastern North America. *Crop Sci.* 34:1097–1100.

Vroomen, H. and H. Taylor. 1992. Fertilizer use and price statistics, 1960–1991. *Statistical Bul.* p. 842. Economics Research Service, U.S. Dept. Agriculture.

Wacquant, J.P., M. Ouknider and P. Jacquard. 1989. Evidence for a periodic excretion of nitrogen by roots of grass-legume associations. *Plant and Soil.* 116:57–68.

Zohary, M. and D. Heller. 1984. *The Genus Trifolium.* Israel Acadamecy of Sciences and Humanities. Jerusalem, Israel.

12. Breeding Methodology

This chapter deals with the various methods used in handling breeding populations in diploid and tetraploid populations of red clover. Aspects of tetraploid clover that are different are discussed in Chapter 13, Tetraploid Red Clover. Genetics and the specific aspects of improving red clover characteristics will be delt with in other chapters.

1. Objectives

In final analysis, most plant breeding objectives are concerned with increased yield, or stability of yields. This includes forage, seed, quality, and pest resistance. It is rather important in choosing objectives to reach a compromise between that which is possible to achieve, and that which is practical. For example, greater persistence for red clover is a desirable objective, but improvement in quality may be a more attainable goal when one considers the resources that must be committed to attain either goal.

Component breeding, or the breeding for the components that affect yield such as leaf area has received little success perhaps because of compensating effects. Breeding for general adaptability (the ability of a population to persist, and produce satisfactory yields in a given area) probably should be the primary goal of most new breeding programs.

As with most cross pollinated forage crops, breeders should keep the following points in mind.

1. Although red clover is difficult to self, it is easily inbred. Close line breeding or sibcrossing results in severe inbreeding depression. Inbreeding by selfing usually cannot be continued beyond two or three generations because of loss of vigor;
2. Inasmuch as seed of most cultivars are produced outside the area of forage production (western states in U.S.), it is necessary to evaluate breeding materials both in the area of forage use and the area of seed production;
3. Most methods used in red clover breeding were first elucidated for corn (*Zea mays*), and utilize spaced plant field populations. The red clover

breeder may be easily led astray by such methods because performance in spaced plant nurseries *will* be different from the broadcast or drilled plantings that farmers use;

4. The performance of individual red clover plants is not nearly as important as performance of their progeny, and some form of progeny testing is a necessity;
5. The importance of breeding red clover populations rather than individual plants probably cannot be over-emphasized. Populations must be kept large to avoid inbreeding depression;
6. Because of the length of life of red clover, a cycle of selection may involve three years, and breeding progress will be slow. If at all possible, schemes should be devised to conduct cycles on an annual or biannual basis, keeping in mind that such schemes may shorten the persistence of the population being selected;
7. Red clover, like many other forage crops, is usually sown in mixtures with grasses and ultimately will have to be evaluated in mixtures, thus slowing the evaluation process. However, early cycles of breeding usually should be conducted in monoculture because the companion crop in the mixture will tend to obscure selection differentials;
8 Red clover is occasionally marketed as hay but usually is marketed through livestock, which adds to the complexity of cultivar evaluation.

2. Sources of genetic variation

In a red clover breeding program that is being initiated in a particular area, the first sources of genetic variation that should be examined are local populations, or recently developed cultivars. These populations are likely to show general adaption to agents affecting production and may possess natural resistance to insects and diseases of the area of utilization. This has been the source of variation for many of the earlier-developed cultivars in the United States and doubtless many other countries.

These populations, or mixtures of them, will be extremely heterogenous and heterozygous, and it is usually a simple matter, with little effort and expense to increase by selection the proportion of plants that have desired characters. If desired levels of characters are not readily apparent in the local population, it still may be possible by various selection methods to increase the gene frequency of a character and reveal hidden variation. An example of this type is the multiple-parted head character in red clover in which the ultimately selected form was much more extreme than that observed in the parental population (Taylor et al., 1985). Other examples could be cited all of which serve to indicate the extreme plasticity of cross pollinated diploid red clover, and indicate that progress is not dependent on high initial variation of the trait being selected (Taylor et al., 1990).

If the desired character cannot be found among locally adapted populations, the next type of population to examine are those from other parts of the country, or even in other countries. In the U.S., seeds of red clover introductions are available from the Regional Plant Introduction Station at Pullman, Washington. An examination of a group of introductions will give a breeder some insight concerning the variability in the species. However, it must be remembered that considerable breeding effort will be necessary to incorporate the desired character into a genetic background that is adapted to any particular area of production and use. Efforts are being made to produce and evaluate "core" collections of red clover so that breeders can make rapid choice of materials for genetic enhancement. (See chapter 16, Germplasm Acquisition, Maintenance and Evaluation).

3. Mutagenesis

Red clover, being a cross pollinated self incompatible species is not ideally suited for most mutagenizing techniques. No suitable method such as in self pollinated crops are available for uncovering changes in recessive genes. Although red clover may be inbred by use of heat, seed set is limited, and not all genotypes are heat sensitive. Another possibility is to use the S_F gene which allows selfing (Taylor, 1982a). Inasmuch as most red clover populations are heterogeneous and each plant is heterozygous, large populations must be used to distinguished between induced and natural variation.

4. Ionizing irradiation

Use of this method for inducing variation in red clover has fallen out of favor, perhaps because of the difficulties mentioned above, and few reports have been seen in the literature during the last 10 years. Previous to this, types of mutagen applied to dry red clover seeds included 30 to 150 k R of gamma radiation, 40 to 100 k R of X-rays, and 1 to 8 hours of thermal neutrons (Taylor et al., 1961). X-ray treatments of 5 to 20 k R have been applied to wet seeds, and to pollen and ovules. Irradiation at lower dosages may have a stimulating effect on seedling emergence (Vanco, 1991). Mutants that have been produced by X-rays on red clover include a short-corolla type, a multifoliolate leaflet type, altered histone content, earliness, greater productivity, and resistance to *Erysiphe* (Taylor and Smith, 1979). Most of these variant types, including shortened corolla tube, generally associated with low seed yields (Goral and Spiss, 1992) have been observed in conventional breeding lines.

5. Chemical mutagens

Seeds of two red clover cultivars were soaked for 18 hours in various concentrations of 3 mutagens by Novoselov (1984). The highest mutation rates (3.3 to 7.7%) were obtained with 0.01 to 0.015% N-nitroso-N-methylurea, 0.03 to 0.05% dimethylsulphate, and 0.05 to 0.01% ethylenenine. From the M2, 92 forms were selected that had useful characteristics including earliness, high inflorescence number, and high dry matter and forage yield. Of the 92 forms, 2 were sterile, and 9 had reduced fertility.

6. Somaclonal variation

The variation found among plants derived from cell culture has been termed "somaclonal variation". These changes have been hypothesized to be the result of mutagens in the culture medium that give rise to the accumulation of point mutations or gross chromosomal changes. Red clover somaclones selected by Bagley and Taylor (1987) under low phosphorus concentration did not express efficiency of phosphorus uptake as whole plants. Most of the somaclones had abnormal leaf arrangements and were low in pollen viability. Somatic chromosome numbers were normal (2n = 14) but meiosis was abnormal. In Russia, *in vitro* selection for resistance to *Fusarium* has been reported (Mazin et al., 1989). The use of somaclones as a source of variability for red clover has the same limitations as ionizing radiations; it is usually not possible to advance the mutant types by selfing.

Two other sources of variability for red clover breeding; interspecific hybridization and induced polyploidy will be considered in other chapters.

7. Population development by hybridization

Often, after sources of variability have been identified, it is desirable to combine the genotypes by sexual hybridization or to achieve homozygosity by selfing. Asexual techniques, including transformation, and somatic fusion will be considered in Chapter 14. Procedures for artificial hybridization and selfing are outlined below.

8. Manual crossing

Equipment used for crossing and selfing include forceps, small scissors, magnifying glasses (10X are satisfactory), small paper or plastic tags, bamboo or plastic stakes, and plant ties. For pollen transfer, toothpicks to which a piece of fine black emery paper has been glued, a sharp pointed lead pencil, or a small folded card are satisfactory. Most of this equipment may be obtained from florist supply companies.

Before flowering, the female parent may be prepared for crossing in the field by using bags of fine muslin with a drawstring closure to protect against pollinating insects. In a screened greenhouse, bags are not necessary. Heads may be supported by stakes to which the stems near the heads are attached. Heads may be trimmed to 15 or 20 recently opened flowers for ease of manipulation.

Emasculation is generally not required for red clover due to its self incompatibility. If in doubt for certain genetic stocks, pollen may be killed by emerging heads in 66.5% ethanol for 10 to 20 seconds and rinsing with water (Bassiri and Smith, 1972).

The optimum stage to pollinate red clover is shortly after the flowers open, and before they begin to wilt which will probably occur in about three days under normal circumstances. For normal pollination, pollen is removed from staminate flowers by inserting the pollinating instrument between the keel and standard petals and applying downward pressure. This causes the staminal column to touch the toothpick. After pressure is removed the staminal column will return to its original position inside the keel. Pollen should be moist, but not wet and should be yellow rather than dry and white. Usually only 5 to 10 flowers can be pollinated with one toothpick load due to dilution of pollen from unemasculated plants used as females. For reciprocal crosses, pollen may be applied alternatively between paired heads of different plants using the same pollinating instrument. Flowers should be kept dry for at least a day or two to prevent aborting of introduced pollen.

After all flowers of a head are pollinated, parentage, date and number of flowers crossed are recorded on a small tag which is looped over the head. It is best to record this information in pencil to prevent loss of information when heads are watered. At this time, the same information should be recorded in a permanent record book where the number of seeds produced also may be listed at a later date. Hands and pollinating instruments should be washed in alcohol before proceeding to the next cross. If toothpicks are used, they may be kept in two small vials, one for those recently used, and ones which have been washed and dried for several days. Heads may be covered with small cotton or plastic bags to protect against pollinating insects.

Seeds of red clover may be produced on excised stems as a convenience for crossing at different locations. Stems with unopened flowers may be severed just above the crown and inserted in vials containing 2% sucrose solution. After the flowers are fully opened, they may be pollinated. Seed should be mature in about two weeks, and dry sufficiently to harvest by 18 to 21 days. Under field conditions, four to six weeks should be allowed for complete maturation of all heads, depending upon climatic conditions.

Fig. 12. 1. Cages used for control of bee pollination.

8.1. *Selfing*

The self incompatible mechanism may be partially overcome in red clover by exposing the heads to a temperature of about 38 °C when flowers are opening. A convenient way to do this is to insert heads through holes into a controlled temperature chamber when flower color is barely showing on unopened heads. After heads fully develop in the chamber, they are then removed to ambient (about 25 °C) temperatures where selfs are made. An alternate procedure is to excise stems with heads in a nutrient solution in glass bottles and maintain them in a controlled environment chamber as described above. Another method tried was selfing *in vitro* but seed set was not increased in red clover (Leduc et al., 1990).

Normally self-incompatible red clover plants may be self-pollinated by gently rubbing the heads between the thumb and fingers, or by toothpick tripping as in the crossing procedure except pollen is transferred only among flowers on the same head. The latter method usually produces the most self seeds.

8.2. *Bee Crossing*

For controlled hybridization with bees, cages or greenhouses may be used (Fig. 12.1). Honey bees (*Apis mellifera* L.) are usually maintained in hives in an apiary until plants which have previously been caged are nearly ready to bloom. Any flowers that may have already opened should be removed to prevent contamination from pollen carried by bees when they are moved into the cages. Honey bees may be moved preferably at night when most of the bees are in the hives. Hives should be strong because most of the free flying bees will not adjust to the caged condition and will die. The newly hatched bees, however, should begin to pollinate the clover heads as they flower. Bees should be watered and fed a 1:1 (v:v) solution of white sugar and water, or a commercially prepared bee food as needed. Red clover in small cages in Kentucky have yielded up to 560 kg of seeds per hectare. This is from rather close spacing. It should be kept in mind that bees are capable of pollinating through the mesh screen if heads grow in close proximity. Weeds in the cages may be controlled by use of herbicides or mulches.

Honey bees have been used to accelerate breeding materials by pollination in the off season in a greenhouse (Zar'yanova, 1989). According to the author, seed set per head was higher than with manual pollination both in diploids and tetraploids. The breeding process was shortened by about two years.

Red clover also may be pollinated by bumble bees (*Bombus* spp.). A technique used in Germany is as follows: *Bombus* spp. queens that have not begun to build a nest, and hence do not collect pollen, are captured in April to develop breeding populations. Bees may be maintained in a refrigerator (5C) for several hours to examine for mite infestation and then used for pollination in greenhouses or field cages (Weber, 1990). Also, it is possible to excavate complete bee nests from the soil and transport them to a cage, but it is probably easier to capture wild bees by swiftly placing a mason jar or other container over them while they are pollinating. As a precaution, contamination may be prevented by capturing bees from sources of pollen other than red clover. The bees are thoroughly washed to remove and abort pollen by adding warm water to the capturing jar and shaking it. In sunny weather, bees may live 10 days to 2 weeks and usually will begin pollinating as soon as they are dry. For control of pollination without cages, great isolation distances are required, because bees have been known to fly several kilometers. Small number of plants may be heavily contaminated by large clover plantings even when isolated up to 450 m. In the converse situation, 450 m may provide sufficient isolation (98 to 99% purity).

9. Seed development, harvest, and storage

Pollinated red clover flowers begin to wilt in about 2 days, but non-pollinated heads will remain receptive up to 10 days depending upon drought and humidity. Clover seeds are physiologically mature about $2\frac{1}{2}$ weeks after pollination but need an extra week or so for drying. Harvested heads may be dried for a few days in a dryer at temperatures not exceeding 45 °C. Equipment used for threshing include corrugated rubber mats for quantities consisting of only a few heads, the Forsberg seed scarifier for somewhat larger quantities, and for still larger amounts, a 90-cm belt thresher. A small stationary combine may be used for larger quantities. For cleaning, the clipper cleaner (Model 2-B)

Fig. 12. 2. Interior of bee crossing cage.

and larger 4 to 6 screen models may be used. A South Dakota seed blower is useful for final cleaning of small lots.

Dry red clover seeds may be stored at 0 to −15 °C without loss of viability. Small amounts of seed in paper envelopes lose little viability when stored in a freezer. Automatic defrosting is not recommended but storage of seed envelopes in moisture proof containers will prevent wetting in case of accidental defrosting. A freezer alarm system also is recommended.

10. Breeding procedures

10.1. *Mass selection*

For red clover breeding, mass selection is a very simple yet effective procedure. It probably should be the first breeding method employed in any new program. An advantage is that large bulk populations often are sown in broadcast or drilled situations so there is little possibility of inbreeding or inadvertent selection for non-useful characters. Usually, seeds are bulked from sources of known value, and are sown in an environment similar to that where the cultivar is to be used. Epiphytotics of diseases, insects, nematodes, or other pathogens that are common to the area may be encouraged by use of coverings, mists, or cages, and by artificial inoculation. Stands may be maintained several years to allow pathogens to aid in selection for longevity. Seed is harvested from old stands where pollination is among selected plants only. Increased general adaptation and pathogen resistance may result in greater yield via increased longevity. Mass selection may be conducted as a single cycle (Vanco, 1989) or by repeated cycles (Balyan and Balyan, 1988, and

Fig. 12. 3. Selecting plants in a spaced plant nursery.

Choo et al., 1994). Mass selection may not be effective for characters with low heritability.

10.2. *Recurrent phenotypic selection*

This process as used for red clover, is very similar to cyclic mass selection. The primary difference is that seeds of half-sib families are maintained individually rather than bulked. The variance among half-sib families allows estimates of narrow sense heritability for the characters under selection. Similar to mass selection, progenies are not tested, which implies that the method will be most effective for simply inherited characters. However, the process may be effective for quantitatively inherited characters that are easily identified or recognized. Escapes or selection of plants with misidentified characters lowers the efficiency of the program, but these will likely be corrected in the next cycle of selection.

Generally, phenotype selection as used for red clover involves identification of plants that often are spaced in a nursery for easy examination (Fig. 12.3). This was the method used in selection for persistence (Montpetit and Coulman, 1991), stem length (Bowley et al., 1984), multiple-parted heads (Taylor et al., 1985), and for plant regeneration in tissue culture (Quesenberry and Smith, 1993). Because red clover flowers at least twice a year, plants may be selected at first bloom, cut back, dug, and moved to a cage for isolation during seed production. Alternatively, if the nursery is isolated from other red clover, non-selected plants may be eliminated, and the selected plants allowed to intercross. These procedures allow the harvesting of seed from half-sib families which may be kept separate and identified in the next and succeeding cycles of selection. By restricting selection to equal numbers of

Spaced-plant nursery

	1	2	3	4	5	6	7	8	9	10	11	12	13	14	15
A	O	O	O	O	O	O	O	O	O	O	O	O	O	O	X
B	O	O	O	O	O	X	O	O	O	O	O	O	O	O	O
C	O	O	O	X	O	O	O	X	O	O	O	O	O	O	O
D	X	O	O	O	O	O	O	O	O	O	O	O	X	O	O
E	O	O	X	O	O	O	O	O	O	X	O	O	O	O	O
F	O	X	O	O	O	O	O	O	X	O	O	X	O	X	O
G	O	O	O	O	X	O	X	O	O	O	O	O	O	O	O
H	O	O	O	O	O	O	O	O	O	O	O	O	O	O	O
I	O	O	O	O	O	O	O	O	O	O	O	O	O	O	O
J	O	O	O	O	O	O	O	O	O	O	X	O	O	O	O

Cycle 1 — Select one plant at first bloom per 10 plant row (10% selection intensity). Other plants killed or removed.

Cage or allow bee crossing for half sib seed production.

Harvest seed from selected plants 1D, 2F, 3E, 4C, 5G, 6B, etc.

Cycle 2 — Germinate seed in greenhouse and start 2nd cycle nursery.

Greenhouse flats

1D	2F	3E	4C	5G	6B	7G	8C	9F	10E	11J	12F	13D	14F	15A

2nd cycle nursery (randomized) †

2F	3E	5G	9F	14F	10E	8C	1D	4C	11J	15A	7G	6B	12F	13D	Rep1
O	O	O	O	O	O	O	O	O	O	O	O	O	O	O	Rep 2
O	O	O	O	O	O	O	O	O	O	O	O	O	O	O	Rep 3
O	O	O	O	O	O	O	O	O	O	O	O	O	O	O	Rep 4

† Each family will be represented by several plants

Cycle 3 — As before.

Fig. 12. 4. Example of phenotypic recurrent selection.

plants from each half-sib family and use of large populations, narrow line breeding and consequent inbreeding depression may be avoided. The method is illustrated in Fig. 12.4.

Also, it may be desirable to impose other restrictions in the selection process if environmental affects are thought to be large. Typically, the most superior plant of each 20 spaced plants may be selected regardless of its performance relative to other plants in the nursery. Where initial variation is rather low, it may be desirable to select a low intensity in early cycles to allow genetic recombination. Selection may be intensified as variation permits.

Many variations of recurrent phenotypic selection are possible. For example, selection may be based on halfsib family means, as was conducted by Taylor et al. (1989) in selecting for 2,4-D tolerance, or a combination of best plant from best family which has been termed geno-phenotypic selection (Taylor, 1987).

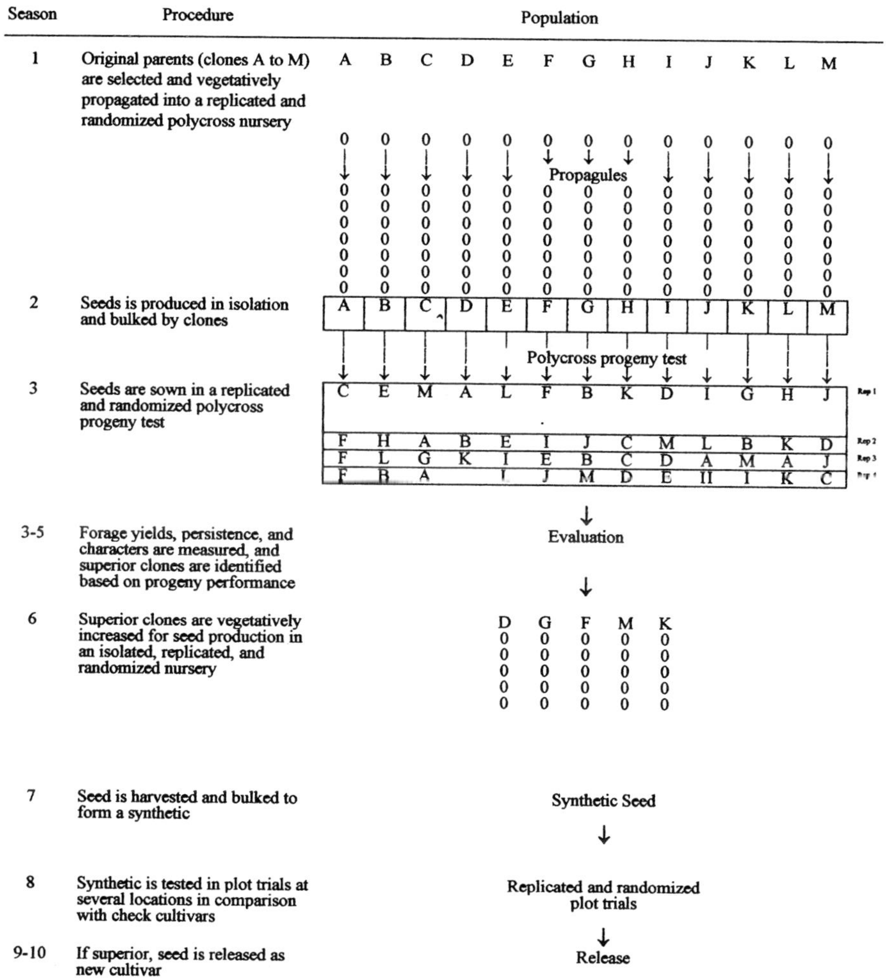

Fig. 12. 5. Polycross method as used in the development of 'Kenstar' red clover.

10.3. *Polycross method*

The objective of the polycross method is to isolate clones with superior general combining ability (GCA) that can be crossed to form an improved synthetic cultivar. This method has been used with some success in red clover. Selected plants previously identified in a polycross nursery or in other breeding procedures are vegetatively increased by crown bud cuttings, rooted in moist vermiculite, sand, peatmoss, or mixtures of these media with soil (Taylor et al., 1962). The procedure follows the diagram provided in Fig. 12.5. Ideally, polycross progeny testing should be conducted in several locations to sample the environment where the cultivar is expected to be used.

Polycross progeny testing also should be by methods that approximate farmer usage of clover, i.e. broadcast or drilled plots. Data from broadcast plots may or may not agree with that obtained from spaced plants. The reliability of different tests of progenies of red clover at three locations was examined by Uzik (1987). Variation for locations was greater than for type of progeny test; however, it was concluded that testing of progeny in spaced plants and in microplots was inferior to row seedings.

After superior clones are identified, breeder seed is produced by intercrossing the selected clones, and if inadequate, for several increases thereafter. When breeder seed is depleted, additional seed must be produced from the original selected clones. Examples of cultivars produced by this method are Kenstar (Taylor and Anderson, 1973), which is maintained with 10 parental clones, the Russian diploid cultivar VIK 7, and the tetraploid cultivar Salynt (Novoselova et al., 1983). The polycross method or a modification of this method described here was used by Godunov (1984) to isolate forms with high general combining ability for green matter yield that were later combined into synthetic populations.

10.3.1. *Advantages of the method*

1. Selection can be conducted for complex characters such as yield or winter hardiness based on progeny testing.
2. The program may be used as the final stage of other breeding programs.
3. Information on general combining ability of clones is determined which may be useful for other types of releases; i.e., double or single cross hybrids.

10.3.2. *Disadvantages of the method*

1. The procedure is very time consuming. For example, in red clover, selection of various resistant clones that can be propagated requires at least 2 years, the polycross nursery requires 2 years; and the polycross progeny test requires 3 additional years. During this time only one cycle of selection has been conducted.

2. The procedure is expensive. Extensive greenhouse space is required to root cuttings and considerable isolated field space is required for the polycross nursery. In addition, progeny testing as few as 100 progenies in field plots at one or more locations is too laborious and expensive for most breeding programs.
3. The usual testing process may not be completely reliable. Progenies of clones usually are tested in single progeny plots where only intraprogeny competition occurs. When selected clones are intercrossed to produce a synthetic cultivar, which is then sown, both intra- and interprogeny competition become operative. Taylor and Kendall (1965) evaluated four polycross progenies, the seeds of which were sown mixed and separately, and concluded that the separate sowing of progenies was likely to give invalid results. They advocated the use of a further mixed progeny test to evaluate differences in competitive ability. General agreement with this concept was illustrated by the research of Savatti (1983) in which advantageous and disadvantageous changes occurred among competitive populations of a synthetic cultivar.
4. Finally, because of virus infection, long-term maintenance of red clover clones is difficult, and maintenance of the cultivar is jeopardized.

10.4. *Strain Building*

This procedure is similar to the polycross progeny test except that parents are maintained by seed populations rather than by clones (Fig. 12.6) (Smith et al., 1973). Populations are developed and maintained separately, and after testing the most superior populations are combined to form a synthetic. The synthetic may be reconstituted as necessary from seed of the parental population maintained in cold storage. Seed of the parental populations may be increased in isolation as necessary.

A form of strain building conducted in France at INRA has been described by Mousset-Declas (1992). In this method, plants in a spaced plant nursery are observed and evaluated for three years. The best plants of the best families are chosen for sib-mating by handcrossing. The next cycle is grown in a nursery and the cycle is repeated for 3 to 5 generations, after which the superior plants (inbred and rather homozygous) of superior families are propagated by cuttings. These clones are established in family mating blocks for seed production under isolation. Some cuttings can be used to obtain tetraploid plants by nitrous oxide treatment. The breeding scheme is the same as for diploid plants. In the subsequent year, harvested seed is used to establish a polycross mating block. Progenies are tested in field plots for three years and families with the highest general combining ability are used to create a synthetic cultivar. This method combines geno-phenotypic, polycross, and strain building methods (Mousset-Declas, 1992).

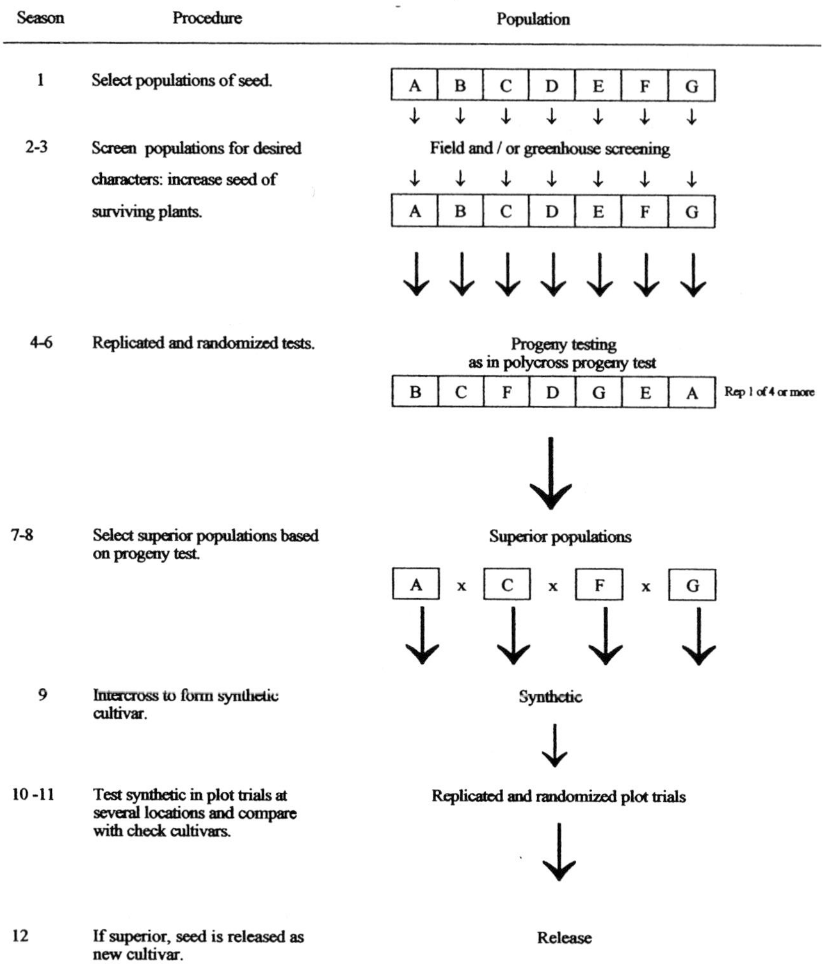

Fig. 12. 6. Strain building as used in the development of "Arlington" red clover.

10.4.1. *Advantages of the method*

1. A very broad genetic base is assured if parents are represented by large numbers of heterogenous and heterozygous genotypes.
2. The procedure is much less laborious because of the maintenance of vegetative clones as in the polycross method is eliminated. Inbred families may be maintained by seeds.
3. It is possible to select clones from these large populations for formation of synthetic cultivars.

10.4.2. *Disadvantages of the method*

1. A relatively long period of time is required to develop, test, and release populations as synthetic cultivars.
2. Because the populations are so broad, genetic gains are likely to be based only on additive genetic variance, and much less on heterotic effects.

10.5. *Backcross procedure*

The method has been used at the University of Kentucky to incorporate resistance to strain 204–1 of bean yellow mosaic virus into the Kenstar cultivar (Taylor et al., 1986). Because the Kenstar cultivar was clonally maintained it was necessary to backcross the source of resistance to each of the 10 parental clones. Since backcrossing to a clone results in inbreeding, the 10 developed populations were intercrossed to regain hybrid vigor. A review of the literature shows no other cultivars of red clover that have been developed by the backcross method.

10.6. *Advantages of the method*

1. The new synthetic cultivar being very similar to its predecessor except for the addition of disease resistance, should require less extensive testing.

10.7. *Disadvantages of the method*

1. In the length of time required to backcross a cultivar thus adding one character, another cultivar probably could be selected that was improved for several characters.
2. For clonally maintained cultivars, backcrossing results in inbreeding and subsequent depression in vigor that must be overcome by appropriate crosses.

10.8. *Single and double-cross hybrids*

One method of producing hybrid cultivars is to produce inbred lines that have homozygous S-alleles thus allowing control of crossing (Taylor, 1982b). Fortunately, only one generation of inbreeding (via selfing) is necessary, because further selfing reduces vigor, fertility, viability of pollen and stability of microsporogenesis (Goral and Spiss, 1986).

Hybrid cultivars produced utilizing the S-allele system exploit both general and specific combining ability that exists in red clover (Anderson et al., 1974; Cornelius et al., 1977). Thus it is possible to capitalize on both additive and dominance genetic variance to produce higher yield and more persistent cultivars than would theoretically be possible by other methods. Two types of crosses with the S-allele system are possible: double crosses in which the

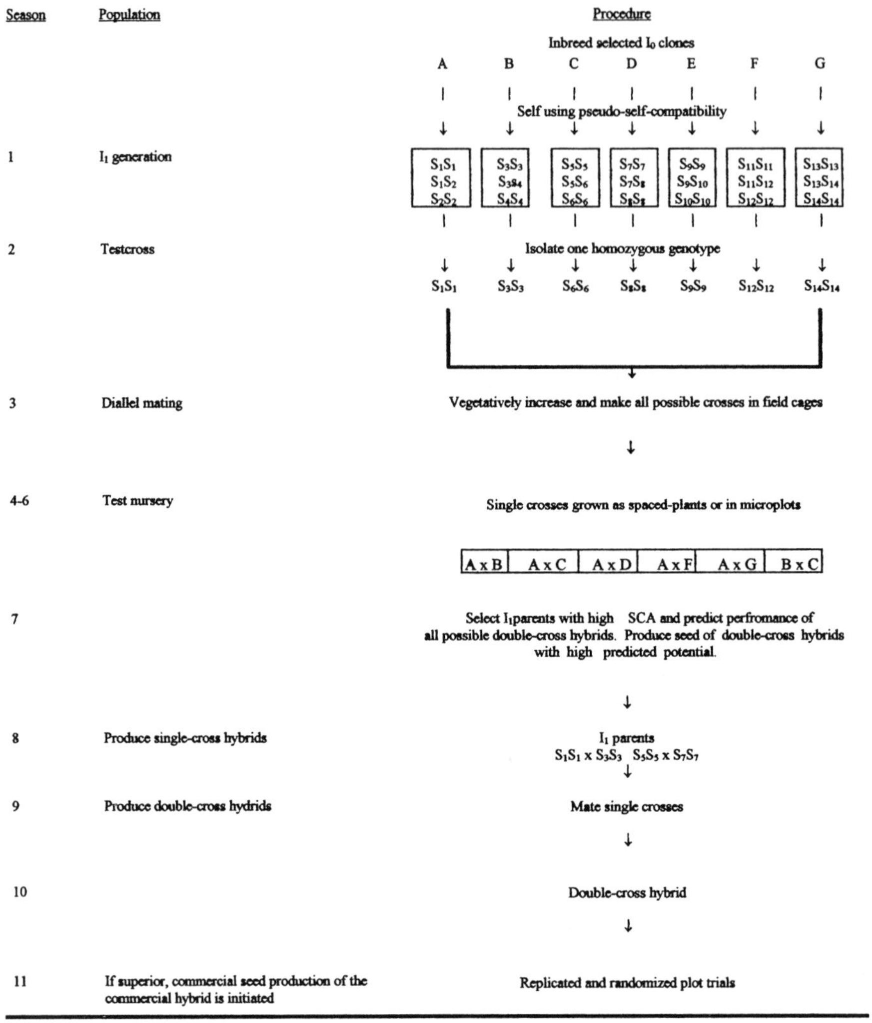

Fig. 12. 7. Method of producing double cross hybrid red clover cultivars using the S-allele system to control crossing.

parental lines are maintained vegetatively on inbred clones (Fig. 12.7), and single crosses in which the parental lines are maintained by alternate selfing and sibbing. S-allele control in both types of hybrids has been shown by Anderson et al. (1972), and Taylor and Anderson (1980). Unfortunately, the double cross hybrids that were released did not exceed the Kenstar cultivar in yield and persistence.

Genetic male steriles have been isolated but are not particularly useful for control of crossing in red clover (Taylor and Smith, 1979). Cytoplasmic male sterility and fertility restoring genotypes were described by Spiss and Goral (1990). Screening is underway to identify stamenless plants that set as many seeds as did the male fertiles.

Hybrid red clover has great potential, but the economic benefit must justify the expense necessary to isolate the rare lines or clones with high specific combining ability. Seed production costs, also expected to be high, must be offset by greater economic benefit as a result of higher yields.

10.9. *Comparison of alternative breeding methods*

The authors are not aware of red clover breeders who have conducted simultaneous programs to be compared for efficiency. Most programs are conducted in the cheapest, most efficient method possible which often does not permit the collection of useful genetic information such as heritability and number of genes conditioning characters.

This limitation accounts for the continued use of mass selection (with control of pollination) and recurrent phenotypic selection, when more sophisticated procedures are available. Recurrent phenotypic selection is a powerful tool because of the large number of selection cycles and the consequent changes that can be induced in relatively few seasons. For this reason phenotypic selection has become and probably will be the method of choice for the future.

Caveats include:

1. Populations must be kept large enough and include sufficient half-sib families to avoid narrow-line breeding and attendant inbreeding depression. This means at least two or three hundred individuals in red clover for each cycle of selection.
2. Characters selected on a spaced plant basis may respond differently in the broadcast sward.
3. Changes in morphological characters induced by selection may be accompanied by undesirable features. An example is the association of multiple-parted heads with fasciated stems (Taylor et al., 1985).

11. Field plot techniques for genotype evaluation

Evaluation procedures for red clover are not markedly different from those that have been described for other small seeded legumes and grasses (see Taylor, 1987). These include: seed bed preparation, lay out of plots, harvesting plots, and large plot testing in several locations (Fig. 12.8), all of which are necessary prior to release for general farmer use. Depending upon the breeding objectives, other types of tests may be conducted. To test quality factors, for example, various kinds of animal grazing or feeding trials may be required. In addition, cultivars that have been changed by genetic transformation may

Fig. 12. 8. Harvesting large forage evaluation plots at INRA, Dijon, France.

be subject to very stringent evaluation by government agencies prior to their release to the public.

References

Anderson, M.K., N.L. Taylor and R.R. Hill, Jr. 1974. Combining ability in I_O single crosses of red clover. *Crop Sci.* 14:417–419.

Anderson, M.K., N.L. Taylor and R. Kirthavip. 1972. Development and performance of double-cross hybrid red clover. *Crop Sci.* 12:240–242.

Bagley, P.C. and N.L. Taylor. 1987. Evaluation of phosphorous efficiency in somaclones of red clover. *Iowa St. J. Res.* 61:459–480.

Balyan, V.P. and G.A. Balyan. 1988. High yielding persistent red clover variety. *Selektsiya i Semenovodstvo, Moscow.* 4:38–39.

Bassiri, A. and R.R. Smith. 1972. Emasculation of self compatible red clover clones with ethanol. *Canad. J. Plant Sci.* 52:846–848.

Bowley, S.R., N.L. Taylor and P.L. Cornelius. 1984. Phenotypic recurrent selection for stem length in 'Kenstar' red clover. *Crop Sci.* 24:578–582.

Choo, T.M., B.R. Christie, G.F. Belanger, Y.A. Papadopoulos and S. Kilyanek. 1994. AC Charlie red clover. *Canad. J. Plant Sci.* 74:825–826.

Cornelius, P.L., N.L. Taylor and M.K. Anderson. 1977. Combining ability in I_1 single crosses of red clover. *Crop Sci.* 17:709–713.

Godunov, I.A. 1984. A method for obtaining combining ability in red clover. *Selektsiyai Semenovodstvo, USSR.* 9:10–12.

Goral, H. and L. Spiss. 1992. Characteristics of red clover forms with a short corolla tube. *Bwletyn Instytutu Hodowii i Aklimatyzacji Roslin.* 183:209–217.

Goral, H. and L. Spiss. 1986. Fertility of red clover inbred lines. *Hodowla Roslin Aklim-atyzacja i Nasiennictwo.* 30:37–46.

Leduc, N., G.C. Douglas, M. Monnier and V. Connolly. 1990. Pollination *in vitro*: effects on the growth of pollen tubes, seed set and gametophytic self-incompatibility in *Trifolium pratense* and *T. repens*. *Theor. Appl. Genet.* 80:657–664.

Liang, G.H.L. 1965. Ph.D. thesis. Univ. Wisconsin Dissertation Absr. 25:6173.

Mazin, V.V., L.I. Lapotyshkina, P.D. Solozhentsev and N.V. Sharapov. 1989. Biotechnological means of obtaining useful forms of clover and lucerne. *Sbornik Nauchnykh Trudov, Vsesoyuznyi Nauchno Issledovatel'skii Institut Kormov.* 42:66–74.

Montpetit, J.M. and B.E. Coulman. 1991. Responses to divergent selection for adventitious root growth in red clover (*Trifolium pratense* L.) *Euphytica.* 58:119–127.

Mousset-DeClas, C. 1992. Red clover breeding in France: From the laboratory to the farm. *Proc. 12th Trifolium Conf. Gainesville.* FL (USA) Mar. 25–27, p 26–30.

Novoselov, M.Y. 1984. Chemical mutagenesis in the production of initial material of red clover. *Selektsiyai Semenovodstvo. USSR.* 6:16–18.

Novoselova, A.S., R.G. Piskovatskaya and V.S. Malashenko. 1983. Results and prospects of using the polycross method in clover breeding. *Sel'skokhozyaistvennaya Biologiya.* 9:23–26.

Quesenberry, K.H. and R.R. Smith. 1993. Recurrent selection for plant regeneration from red clover tissue culture. *Crop Sci.* 33:585–589.

Savatti, M.M. 1983. Intraspecific competition in red clover and its breeding implications. *Probleme de Genetica Teoretica si Aplicata.* 15:277–284.

Smith, R.R., D.P. Maxwell, E.W. Hanson and W.K. Smith. 1973. Registration of Arlington red clover (*Trifolium pratense* L.) *Crop Sci.* 13:771

Spiss, L. and H. Goral. 1990. Characteristics of red clover plants containing male-sterile cytoplasm and an attempt to find sterility maintaining lines. *Hodowla Roslin Aklimatyzacji i Nasiennictwo.* 30:27–36.

Taylor, N.L. 1982a. Registration of gene marker germplasm for red clover (Reg. No. GP1 to GP11). *Crop Sci.* 22:1269.

Taylor, N.L. 1982b. Stability of S alleles in a doublecross hybrid of red clover. *Crop Sci.* 22:1222–1225.

Taylor, N.L. 1987. Forage Legumes. pp. 209–248. In Fehr, W.R. (ed.), *Principles of Cultivar Development.* Vol. 2. MacMillan Publ. Co. N.Y.

Taylor, N.L. and M.K. Anderson. 1973. Registration of Kenstar red clover. *Crop Sci.* 13:772.

Taylor, N.L. and M.K. Anderson. 1980. Maintenance of parental lines for hybrid red clover. *Crop Sci.* 20:367–369.

Taylor, N.L., P.L. Cornelius and M.G. Long. 1985. Phenotypic recurrent selection for multiple-parted heads in red clover. *Crop Sci.* 25:489–494.

Taylor, N.L., S.A. Ghabrial, S. Diachun and P.L. Cornelius. 1986. Inheritance and backcross breeding of the hypersensititve reaction to bean yellow mosaic virus in red clover. *Crop Sci.* 26:68–74.

Taylor, N.L., E. Gray, W.H. Stroube and W.A. Kendall. 1961. Some effects of gamma and neutron irradiation of seeds on germination and seedling growth of red clover. *Crop Sci.* 1:458–460.

Taylor, N.L. and W.A. Kendall. 1965. Intra- and inter-polycross progeny competition in red clover, *Trifolium pratense* L. *Crop Sci.* 5:50–52.

Taylor, N.L. and R.R. Smith. 1979. Red clover breeding and genetics. *Advan. Agron.* 31:125-154.

Taylor, N.L., R.R. Smith and J.A. Anderson. 1990. Selection in red clover for resistance to northern anthracnose. *Crop Sci.* 30:390–393.

Taylor, N.L., W.H. Stroube, W.A. Kendall and E.N. Fergus. 1962. Variation and relation of clonal persistence and seed production in red clover. *Crop Sci.* 2:303–305.

Taylor, S.G., D.D. Baltensperger and K.H. Quesenberry. 1989. Recurrent half-sib family selection for 2,4-D tolerance in red clovers. *Crop Sci.* 29:1109–1114.

Uzik, M. 1987. Reliability of different tests of offspring of red clover. *Vedecke Prace Ustavu Prestanoch Krmoviny.* 21:31–48.

Vanco, B. 1989. Efficacy of a single cycle of mass selection in red clover for resistance to root rot. *Vedecke Prace Vyskumneho Ustavu Rastlinnej Vyroby v Piest'anock Krmoviny.* 22:73-82.

Vanco, B. 1991. Influence of mutagens on certain parameters and properties of red clover seedlings. *Ved Pr Vysk Ustavu Rastl Vyroby Piestanoch Krmoviny.* 23:99–112.

Weber, D. 1990. Practical experiences in breeding bumblebee populations for purposes of pollination in closed areas. *Wiss Z Martin Luther Univ. Halle Wittenberg Math Naturwiss Reihe.* 39:3–7.

Zar'yanova, Z.A. 1989. Accelerated production of red clover breeding material by means of bee pollination in the greenhouse. *Sbornik Nauchnykh Trudov, Vsesoyuznyi Nauchno Issledovatel'skii Institut Kormov.* 42:31–36.

13. Tetraploid Red Clover

Plant breeders have been fascinated with doubling of chromosomes of plant species since the discovery of colchicine in 1937. The early research on doubling of chromosomes of red clover, conducted primarily by Swedish researchers, met with considerable success over the years, and continuing to the present, many tetraploid cultivars have been produced. Plant breeders of other countries, no doubt impressed with the Swedish results, have developed tetraploid cultivars that exceed diploid counterparts in many respects. The successes of the Swedish program was documented by Sjodin and Ellerstrom (1986).

1. Characteristics of tetraploid red clover

1.1. *Self-Fertility*

Doubling of chromosomes upsets the relationships of the S-alleles that control self-fertility. The relationship of the 2 S-alleles in the pollen grains of autoploids may involve dominance, independence and competition. In a program of doubling at the zygotic stage approximately 25% of the S-alleles in pollen grains gave rise to competition which suppressed or depressed the incompatibility reaction and increased self fertility (Picard, 1985) . Up to 3.1% self fertile plants occur in diploid cultivar populations and from 0.8 to 7.2% among tetraploid populations. Apparently such increase in self-fertility does not result in inbreeding depression inasmuch as the subsequently developed tetraploid cultivar exceeded the diploid in several characteristics (Drach et al., 1986).

1.2. *Flowers and seeds*

Tetraploid forms generally are larger than diploids in most plant parts including flowers and seeds. Corolla tubes range from 7.9 to 11.3 mm long rendering pollination by honey bees (*Apis mellifera*) quite difficult (Nikovitz, 1985). The flowers of tetraploids tend to be pollinated mostly by large bumble bee

(*Bombus*) species (Alekseev, 1981). Honey bees will pollinate tetraploid red clover but 12,000 to 13,000 are required to pollinate one ha compared to 6000 bumble bees (Haslbachova and Vesely, 1981). Seeds of tetraploids are heavier than diploids (2.67 to 2.77 g as compared to 1.73 to 2.05 g for 1,000 seeds) (Sevecka, 1987). This results in greater seedling emergence of tetraploid than diploid seeds but may require higher sowing rates due to the lower number of tetraploid seeds per unit of weight as contrasted to diploids.

1.3. *Disease resistance*

Most researchers have shown increased disease resistance to be one of the attributes of tetraploid red clover. For example, increased resistance to *Sclerotinia trifolium*, and *Kabatiella caulivora* have been reported (Yamada, and Hasegawa, 1990) but Arseniuk (1989) reported no increased resistance to *Erysiphe polygoni* in tetraploid as compared to diploid cultivars.

1.4. *Persistence and forage yield*

Generally most European and Japanese researchers report greater persistence and forage yield for tetraploids than for diploids (Tomaszewski, 1989; Sheglov and Antonova, 1983; Jonsson, 1985) whereas the American experience has been the opposite (Taylor and Wiseman, 1985).

An example of the increased forage yield was given by Julen (1971). Increases in the yields of the first harvest ranged up to 115% of that of the diploid cultivar Merkur, and in the second harvest ranged up to 138% of Merkur. According to Sjodin and Ellerstrom (1986) the most successful tetraploid to that time was 'Sally', released in 1974 as a selection from a population with a very wide genetic background (Sjodin, 1981). Its dry matter yields relative to the standard diploid cultivar Bombi was 104 for the first year ley, and 120 for the second year ley. The newer cultivar Fanny may be even higher forage yielding (Sjodin and Ellerstrom, 1986).

1.5. *Seed production*

Low seed yields and high seed prices traditionally have been the primary limiting factors to the widespread use of tetraploid cultivars. Some success has been achieved in overcoming the problem. Seed yield of tetraploids have been increased up to 42.5% by treatment with Alar 85 plus boron (Jakesova and Svetlik, 1986). Seed yields also have been increased by breeding. In Sweden, the tetraploid cultivar Sally produced an average of 72% of the seed yield of a diploid cultivar but Fanny, a new cultivar at that time had seed yields of 12% over Sally (Andersson, 1982). According to Sjodin (personal communication) during 1993–94, tetraploids constituted between 40 to 50% of certified seed production in Sweden. In France, the corresponding figure was 4 to 6% (Mousset-Declas, personal communication). Seed yields equal to those of diploids for the tetraploid 'Radyka' have been produced in Poland (Goral, 1983). However, Tomaszewski (1988, 1989) compared 8 diploid and 7 tetraploid forms in Poland, and found that seed production was 30–60% higher in the diploids than in the tetraploids. Phenotypic recurrent selection was effective in Poland to increase seed yields derived from 2×-4× crosses (Broda et al., 1992). A review of the literature indicates that breeding for seed yield is underway in Sweden, Germany, Romania, Russia, Slovakia, Hungary, Czech, Lithuania, a very limited amount in the U.S.A., and perhaps in other countries. Low seed yields in the U.S. have limited distribution of tetraploid cultivars.

2. Methods of producing tetraploids

2.1. *Colchicine*

The first method used was that involving colchicine, which is still the method of choice in Sweden, and perhaps other countries. The usual method is to apply weak solutions (approximately 0.02 to 0.05%) to germinating seeds either in a vacuum or at ambient atmospheric pressures. Alternatively, solutions (approximately 0.4%) in lanolin pastes may be applied to apical meristems of growing plants (Novoselov et al., 1989). Percentage of plants doubled is usually quite low and frequencies of mixaploid or sectorial chimeras are quite high. Nevertheless, when the amount of plant material to be doubled is not limited (as with seeds), colchicine remains the chemical of choice and has been used for the development of many cultivars (Bilis and Pivoryunas, 1988). A relatively new technique involves in vitro use of colchicine and is particularly well adapted for doubling sterile interspecific hybrids, and is also effective for other vegetative applications (Anderson et al., 1991).

2.2. *Nitrous oxide*

To overcome the difficulty of sectorial chimeras, French researchers (Berthaut, 1968) treated red clover plants with nitrous oxide and obtained a high percentage of doubled plants. Briefly, the technique consists of making crosses and after 24 hours, the plants are enclosed in a gas-tight chamber (Fig. 13.1). Nitrous oxide is introduced into the chamber and maintained at 6 bars atmospheric pressure (approximately 90 p.s.i.). After 24 hours, plants with heads are removed and placed on a greenhouse bench or incubator at about 20 °C for a seed maturation period of about three weeks. Seeds are then rubbed out, and germinated. Plants are grown to flowering and classified as to ploidy level based on dry pollen shape. Haploid pollen is cylindrical and diploid pollen is tetrahedronal (Taylor et al., 1976). The percentage of tetraploids

Fig. 13. 1. Gas tight chamber used for treating red clover with nitrous oxide at INRA.

obtained by Taylor was 71% and by Slesaravichyus and Sliesaravicius (1988) was from 60 to 70%. The technique also has been used in Japan by Matsu-ura et al. (1974) with similar results.

2.3. *Gametic non-reduction*

The third method for doubling chromosome numbers in red clover is gametic restitution or the unreduced gamete method. The use of the method to maximize heterozygosity in potato (*Solanum tuberosum* L.) and alfalfa (*Medicago sativa*) stimulated its use for red clover. Crosses of a diploid with a tetraploid plant may be made in either direction, 2×-4× or 4×-2×. If the diploid plant is used as the female, female gametic non-reduction will result in tetraploid seed; in the other direction, male non-reduction is involved. Once one tetraploid is produced (2×-4×) the reciprocal cross (4×-2×) will result in bilateral sexual polyploidization (BSP) that theoretically should result in the highest level of heterosis (Mendiburu and Peloquin, 1977). Alternatively, male and female 2n gamete-producing plants can be used in a 2×-2× cross and any tetraploid progeny may be bilateral sexual polyploids.

The frequency of 2n (unreduced) female gametes from 2×-4× crosses of red clover from 3 cultivars and one experimental strain with plants from a nitrous oxide derived tetraploid population ranged from 1 in 1500 to 1 in 11,000 flowers (Taylor and Giri, 1983). About half the tetraploids exhibited aneuploidy having 27, 29, or 30 chromosomes as well as the expected 28. In a later study, up to 2.3% triploids were found among progenies of 2×-4× crosses; tetraploid production ranged up to 66.2% of progenies in one cross and averaged 6.2 2n eggs per 1000 flowers (Taylor and Berger, 1989).

For estimating the frequency of 2n male gametes from 4×-2× crosses, a white flowered tetraploid red clover was crossed in a bee cage with plants of a diploid red-flowered population. Seeds were harvested from the white flowered progeny, and were germinated and grown to flowering. Plants that were red flowered (indicating cross progeny) were examined for polyploidy. Most of the red-flowered plants were triploid and may have interfered with tetraploid production, the frequency of which was estimated to be about 1 in 10,000 flowers (Taylor and Wiseman, 1987). In a later study involving different clones, the percentage of tetraploid plants from 4×-2× crosses was 86.7% and for triploids, 13.3% (Taylor and Berger, 1989).

Parrot and Smith (1984) screened 600 diploid plants (6 cultivars) for 2n pollen by examining dry pollen under a microscope. Eighteen plants (3%) produced at least 1% 2n pollen and individual plants ranged from 1 to 84%. One plant produced both 2n eggs and 2n pollen. In later research, three cycles of phenotypic recurrent selection for 2n pollen production resulted in a change from 0.4% in the original population to 47.4% in the third-cycle population (Parrott and Smith, 1986).

Bilateral sexual polyploidization (BSP) was used to tetraploidize red clover in 2×-2× crosses (Parrott and Smith, 1985). Diploid plants that were known producers of 2n eggs were crossed with plants known to produce 2n pollen, and 13% of the progeny in one cross were tetraploid. The female plant produced an estimated 0.14 2n eggs per thousand flowers crossed.

3. Comparison of methods for producing tetraploids

The three basic methods of doubling chromosome numbers; colchicine, nitrous-oxide, and unreduced gametes are all useful and each has its unique applications. For example, colchicine is the only technique of the three that may be applied to sterile tissue, and also is particularly useful for treatment of germinating seeds. Being an asexual method, it may provide different ploidy level copies of an originating plant, enabling study of gene action at different ploidy levels. The percentage of doubled plants may be low but this is not a problem as long as seeds and other tissues are plentiful.

Colchicine application usually results in mixaploid tissues in which an individual plant may consist of diploid, tetraploid, or even higher ploidy tissues. However, random crossing among such plants will produce progenies of different ploidy levels that may be sorted out by pollen or root-tip chromosome examination. Up to half the progeny deriving from colchicine treatment may be aneuploids. Inasmuch as euploids themselves tend to generate aneuploids which in turn are somewhat lower seed yielding than euploids, an equilibrium seems to be reached that is not detrimental to the tetraploid population (Giri et al., 1983). Sjodin and Ellerstrom (1986) illustrated plants at the rosette stage with 26, 27, 28, 29, 30 and 31 chromosomes.

Nitrous-oxide treatment apparently involves doubling the chromosomes of the zygote and thus has the potential of providing doubled versions of F_1 crosses (Giri et al., 1983). Thus, two populations are available for comparison. Although equipment for nitrous-oxide is more expensive than equipment for colchicine application, the cost is not prohibitive. No mixaploids are produced, and the percentage of tetraploidy is much higher than with colchicine, commonly ranging up to 70%. A low percentage of octoploids may be produced. Aneuploids occur in about the same frequency as with colchicine application.

The unreduced gamete or gametic non-reduction method involves doubling of reproductive tissue and is apparently the mechanism most prevalent in nature. The method has the potential to provide maximum heterozygosity because the genomes are unlike, thus diverse in allelic frequency. According to this hypothesis, colchicine doubling being an asexual method would result in less heterozygosity and more inbreeding depression due to less diversity of alleles. If this hypothesis is correct, colchicine-derived tetraploid populations would be expected to have more quadrivalent associations and fewer bivalent associations at metaphase I than either nitrous oxide or unreduced gamete derived populations. Taylor and Giri (1983) showed that the colchicine derived tetraploid cultivar, Radyka, did in fact have about half of its chromosomes pairing as bivalents compared to nitrous oxide and unreduced gamete cultivars (2×-4×) at approximately 75%. However, in a later study, Taylor and Wiseman (1987) found only 55.5% bivalent pairing in a small population developed by 4×-2× crosses. In a bilateral sexual polyploidization (BSP) derived population, only 42.6% bivalent pairing was found. The reasons for the low frequency of bivalent pairing in the 4×-2× and BSP populations are unknown. However, it should be noted that the populations being compared differed in size and origin. The difference also could be caused by genes that control pairing (Tofte and Smith, 1989). Tetraploid germplasms but no cultivars developed by polyploidization have been released in the United States. These germplasms are lower yielding and less persistent than comparable diploids (Taylor, 1983, 1989).

The data indicate that tetraploid populations may be developed by any of these methods provided that large numbers of parents of diverse origin are used to avoid close line breeding and subsequent inbreeding depression. It also is probably advisable to start selection for improved seed production at the diploid level (Sjodin and Ellerstrom, 1986). The reason why tetraploids have been successful in Europe but not in the United States is probably not related to the method of producing tetraploids, but otherwise is not understood. One suggestion is that tetraploids are not as well as adapted to stressful conditions that are prevalent in the United States. This was not supported by the data of Makarov (1983) who found higher dry matter yields of tetraploids than diploids when populations were grown under drought. Likewise, tetraploids have maintained their superiority over diploids in several plant characteristics including plant weight at three different altitudes (Graman and Sakova, 1988) . Plant breeders in Europe have been involved in tetraploid breeding much longer than in the United States, and apparently considerable time is required for the genes of red clover to become adapted to existence in cells of higher ploidy.

References

Alekseev, V.A. 1981. Pollination of tetraploid clover. *Pchelovodstvo.* 6:19–20.

Andersson, B. 1982. Dry matter yield and seed-setting ability in new varieties of herbage legumes. *Nordisk Jorbrugsforskning.* 64:367.

Anderson, J.A., C. Mousset-Declas, E.G. Williams and N.L. Taylor. 1991. An *in vitro* chromosome doubling method for clovers (*Trifolium* spp). *Genome.* 34:1–5.

Arseniuk, E. 1989. Effect of the polyploidization of red clover (*Trifolium pratense* L.) on winter hardiness and resistance to some diseases. *Hodowla Roslin, Aklimatyzacjai Nasiennictwo.* 33:1–40.

Berthaut, J. 1968. L'emploi du, protoxyde d'ozate dans la creation de varieties aulotetraploides chez le trefl violet (*Trifolium pratense* L.). *Ann. Amelior. Plantes.* 18:381–390.

Bilis, Y. and Y. Pivoryunas. 1988. The use of the polyploidy method of breeding early red clover. *Sbornik Nauchnykh Trudov, Litovskii Nauchno Issledovatel'skii Institut Zemledeliya.* 58:12–19.

Broda, Z., J. Nayracola, G. Konieczny, G. Dyba and L. Torz. 1992. In Veronesi, F., S. Bullitta and S. Coredda (eds.), *Proc. 17th Eucarpia Meeting: Ploidy and Chromosome Manipulation in Forage Breeding.* Rome, Italy.

Drach, N.P., V.M. Rogovchenko, T.N. Skorobagat'ko and G.M. Efimenko. 1986. Significance of self pollination in the reproduction of early red clover. *Sbornik Nauchnykh Trudov po Prikladnoi Botanike, Genetike i Selektsii.* 103:31–34.

Giri, N., N.L. Taylor and G.B. Collins. 1983. Chromosome stability and fertility of a nitrous oxide-derived tetraploid population of red clover. *Crop. Sci.* 23:45–48.

Goral, S. 1983. Utilization of *Trifolium pratense* L. ecotypes in the breeding programme of tetraploid cultivar-Radyka. In *The Utilization of Genetic Resources in Fodder Crop Breeding.* Aberystwyth, U.K. 3–16 September, 1982. pp. 275–283.

Graman, J. and L. Sokova. 1988. Modifying influence of different agroeological conditions on cultivar characteristics of the diploid and tetraploid forms of red clover. *Sbornik Vysoka Skola Zemedelska Praha, Agronomicka Fakulta v Ceskych Budejovicich, Fytotechnika Rada.* 5:29–42.

Haslbachova, H. and P. Vesely. 1981. Tetraploid red clover (*Trifolium pratense* L.) pollination. *Acta Universitatis Agriculturae Brno, A Facultlas Agronomica.* 29:155–166.

Jakesova, H. and V. Svetlik. 1986. Application of trace elements with growth regulators in tetraploid red clover varieties. *Sbornik Vedeckych Praci, Osesva,-Vyzkumny a Slechtitelsky Ustav Picninarsky, Troubsko u Brna.* 10:23–31.

Jonsson, H.A. 1985. Red clover (*Trifolium pratense*) Sara. *Agri Hortique Genetica.* 43:43–51.

Julen, G. 1971. Breeding for increased yield of dry matter and of seed in tetraploid red clover. *Proc. Eucarpia Fodder Crop Sect. Meet.,* Lusignan. pp. 157–167.

Makarov, N.M. 1983. On the question of drought resistance of tetraploid *Trifolium pratense*. *Sibirskii Vestnik Sel'skokhozyaistvennoi Nauki*. 4:48–50.

Matsu-Ura, M., Y. Maki and R. Hayokawa. 1974. Inducing autopolyploids of red clover (*Trifolium pratense*) with nitrous oxide. Res. Bull. Hokkaido *Natl. Exp. Stn. No.* 108: 99–105.

Mendiburu, A.O. and S.J. Peloquin. 1977. The significance of 2n gametes in potato breeding. *Theor. Appl. Genet.* 49:53–61.

Nikovitz, A. 1985. Bee pasture value of red clover varieties. *Meheszet.* 33:1,7.

Novoselov, M.Y., K.K. Mazur, N.P. Oshanina, I.N. Ostretsova and L.S. Zarutskaya. 1989. Method of producing mutants and mixoploids of red clover under protected cultivation. USSR-Patent A.S., 1470251.

Parrott, W.A., R.R. Smith and M.M. Smith. 1984. Production of 2n pollen in red clover. *Canad. J. Genet. Cytol.* 27:64–68.

Parrott, W.A. and R.R. Smith. 1985. Bilateral sexual tetraploidization in red clover. *Crop Sci.* 27:64–68.

Parrott, W.A. and R.R. Smith. 1986. Recurrent selection for diploid pollen formation in red clover. *Crop. Sci.* 26:1132–1135.

Picard, J. 1985. A short note about the relations between S-alleles in the dihaploid pollen grains of tetraploid red clover plants. *Plant Cell Incompatibility NewsLetter.* 17:27–28.

Sevecka, L. 1987. Effect of sowing regimes on number and percentage of emerged plants in red clover. *Vedecke Prace Vyskumneho Ustavu Rostlinnej Vyroby v Piest'anoch, Krmoviny.* 21:115–126.

Sheglov, Y.S. and N.E. Antonova. 1983. Tetraploid red clover cv. Temp and pecularities of its cultivation technique. *Selektsiya i Semenovodstvo.* 4:33–35.

Sjodin, J. 1981. Selection for seed setting capacity in tetraploids of clover and grasses. *Proc. Eucarpia Fodder Crop Sect. Meet.* Merelbeke. pp. 163–168.

Sjodin, J. and Ellerstrom. 1986. Autopolyploid forage crops. In Olsson, G. (ed.), Svalof 1886-1986. *Research and Results in Plant Breeding.* Svalof AB, Svalov, Sweden.

Slesaravichyus, A. and A. Sliesaravicius. 1988. Polyploidization of red clover with nitrous oxide. Problemy ekologicheskogo monitoringa i geneticheskie aspekty ornitofauny i drugikh organizmov. 2. Problemy geneticheskogo i ekologicheskogo monitoringa rastenii i zhivotnykh. 1988:52–53. Vilnius, Lithuanian SSR.

Taylor, N.L. 1983. Registration of diploid and derived tetraploid red clover germplasm. *Crop Sci.* 23:1226.

Taylor, N.L. 1989. Registration of tetraploid red clover germplasms from 2×-4× and 4×-2× crosses. *Crop Sci.* 29:1578–1579.

Taylor, N.L., M.K. Anderson, K.H. Quesenberry and Linda Watson. 1976. Doubling the chromosome number of *Trifolium* species using nitrous oxide. *Crop Sci.* 16:516–518.

Taylor, N.L. and S. Berger. 1989. Polyploids from 2×-4× and 4×-2× crosses in red clover. *Crop Sci.* 29:233–235.

Taylor, N.L. and N. Giri. 1983. Frequency and stability of tetraploids from 2×-4× crosses in red clover. *Crop Sci.* 23:1191–1194.

Taylor, N.L. and E.O. Wiseman. 1985. Methodology and breeding of tetraploid red clover. *Proc. Intern. Grassl. Congr.* 15:244–245. Kyoto, Japan.

Taylor, N.L. and E.O. Wiseman. 1987. Triploids and tetraploids from 4×-2× crosses in red clover. *Crop. Sci.* 27:14–18.

Tofte, J.E. and R.R. Smith. 1989. Cytology of bilaterally derived tetraploid red clover. *Crop. Sci.* 29:37–41.

Tomaszewski, Z. Jr. 1988. Comparison of the agronomic value of diploid and tetraploid forms of red clover (*Trifolium pratense* L.). *Biuletyn Instytutu Hodowli i Aklimatyzacji Roslin.* 168:55–64.

Tomaszewski, Z. Jr. 1989. Comparison of seed production in diploid and tetraploid varieties and lines of red clover (*Trifolium pratense* L.). *Biuletyn Instytutu Hodowli i Aklimatyzacji Roslin.* 169:3–11.

Yamada, M. and T. Hasegawa. 1990. New forage crop varieties registered by the Ministry of Agriculture, Forestry and Fisheries in 1989 and 1990. *Jap. J. Plant Breed.* 40:549–554.

14. Tissue Culture

The initial interest in application of plant cell, tissue, and organ culture in red clover arose from attempts by various researchers to create interspecific hybrids between red clover and other *Trifolium* spp. Early research revealed that several of these hybrid combinations initiated development of a hybrid embryo, but the embryo aborted prior to maturity (Chapter 2). Attempts to rescue embryos of red clover × *T. medium, T. alpestre,* or *T. rubens* on the available media were not successful (Evans, 1962). Embryos from crosses of white clover with three other species were successfully rescued on these media. Little additional research was reported on applications of cell, tissue, and organ culture in red clover until the late 1970's. A general review of tissue culture research in the genus *Trifolium* was presented by Rupert and Collins (1985). The present chapter focuses on red clover and recent results. Some review of earlier media development work is presented.

1. Media development

The basic requirements for any type of *in vitro* culture are a basic mineral salts mixture, supplemented with organic compounds including certain amino acids, vitamins, and sugars. Auxin and cytokinin analogs are added in varying ratios to promote or suppress root or shoot development. The primary basal salts mixtures that are used for red clover tissue culture are B5 (Gamborg et al., 1968, 1976), MS (Murashige and Skoog, 1962), L2 (Phillips and Collins, 1979a), and EC6 (Maheswaran and Williams, 1984). A comparison of the quantity of macro- and micro-mineral salts and organic compounds contained in these four media is shown in Table 14.1. The B5 and MS media were developed initially for soybean and tobacco, respectively. The L2 medium of Phillips and Collins (1979a) was optimized specifically for red clover through evaluation of over 125 media variations. These researchers concluded that high levels of 2,4-dichlorophenoxy acetic acid (2,4-D) as the auxin source were generally detrimental to red clover callus growth. The L2 medium has higher levels of Ca and Mg than MS or B5 as well as comparatively high levels of K and P. Levels of several micronutrients are also relatively high.

Table 14.1

Comparison of inorganic mineral composition of four media used in red clover tissue culture.

Macronutrients	B5		L2		MS		EC6	
	mg/l	mM	mg/l	mM	mg/l	mM	mg/l	mM
NH_4NO_3			1000	12.5	1650	20.6	600	7.5
KNO_3	2500	25.0	2100	20.8	1900	18.8	950	9.4
$CaCl_2 \cdot 2H_2O$	150	1.0	600	4.1	440	3.0		
$CaCl_2$							166	1.5
$MgSO_4 \cdot 7H_2O$	250	1.0	435	1.8	370	1.5	185	0.75
KH_2PO_4			325	2.4	170	1.25	170	1.25
$(NH_4)_2SO_4$	134	1.0						
$NaH_2PO_4 \cdot H_2O$	150	1.1	85	0.6				
Micronutrients	mg/l	μM	mg/l	μM	mg/l	μM	mg/l	μM
KI	0.750	4.5	1.0	6.0	0.830	6.0	0.83	6.0
H_3BO_3	3.000	48.5	5.0	80.9	6.200	100.0	0.62	10.0
$MnSO_4 \cdot 4H_2O$					22.300	100.0	2.23	10.0
$MnSO_4 \cdot H_2O$	10.000	60.0	15.0	90.0				
$ZnSO_4 \cdot 7H_2O$	2.000	7.0	5.0	17.4	8.600	30.0	0.86	3.0
$Na_2MoO_4 \cdot 2H_2O$	0.250	1.0	0.4	1.7	0.250	1.0	0.25	1.0
$CuSO_4 \cdot 5H_2O$	0.025	0.1	0.1	0.4	0.025	0.1	0.0025	0.01
$CoCl_2 \cdot 6H_2O$	0.025	0.1	0.1	0.4	0.025	0.1	0.0025	0.01
Na_2EDTA	37.300	100.0	33.6	90.0	37.300	100.0	37.30	100.0
$FeSo_4 \cdot 7H_2O$	27.800	100.0	25.0	90.0	27.800	100.0	27.85	100.0
Sucrose(g/l)	30		25		20		40	

The EC6 medium is lower in levels of several major and minor salts than the other three media. A comparison of the organic components of these four media is presented in Table 14.2. The EC6 medium uses yeast extract rather than defined organics as in the other three media. Recommended levels of growth regulators are discussed in other sections of this chapter, because ratios and kinds of auxins and cytokinins regulate cell and tissue development. Additional media modifications suggested for various tissues and organs also are discussed in other sections of this chapter.

2. Meristem culture

Successful *in vitro* culture of red clover meristems using the basic L2 medium has been reported (Phillips and Collins, 1979b). Explant sources included crown and axillary shoot meristems. Picloram at 4.0 μg/l and benzylamino purine (BA) at 1.0 mg/l were superior for direct reconstitution of shoots without callus growth. Such direct reconstitution is desirable to avoid possible somaclonal variations that may result from callus tissue. Higher levels of picloram stimulated callus proliferation and multiple shoots which might be used for rapid asexual propagation of desirable red clover clones. Plants were produced from both crown and axillary meristems, but plants from the latter appeared to lack the ability to form branches. Crown-derived plants grew normally and developed new crowns.

The primary purpose of the above research was to eliminate viruses from the 10 parental clones of the synthetic cultivar Kenstar. Bioassay of plants produced from meristem culture showed that when the meristem dome with the first foliar primordia (0.1–0.4 mm) was used as explant source, about 85% were symptomless for the virus. When larger explants (0.4–0.7 mm) were used only about 50% of plants were symptomless. Thus, smaller explants put into culture should be used to produce the highest frequency of virus-free plants.

More recently, red clover has been cured of alfalfa mosaic virus (AMV) and of bean yellow mosaic virus (BYMV) using meristem culture (Servitova and Pokorny, 1987). In one pass through meristem culture 51% of recovered plants were free of AMV, 76% were free of BYMV, and 38% were free of both viruses. Additional research at the Welch Plant Breeding Station has used shoot tip culture to eliminate white clover mosaic virus (WCMV) and red clover necrotic mosaic virus (RCNMV) from red clover (Dale and Cheyne, 1993). These workers used shoot tips ranging in size from <0.6 mm up to 3.0 mm in length. Their results were similar to those of Phillips and Collins (1979b) in that efficiency of virus elimination was 100% for plants regenerated from tips <0.6 mm length but only 12% for plants regenerated from tips larger than 1.9 mm. Conversely, efficiency of plant regeneration from the <0.6 mm shoot tip explants was 30% vs 76% for shoot tips >2.5 mm in length. Thus, one must balance percentage virus elimination against percentage regeneration when using meristem culture to eliminate viruses or other systemic organisms from red clover.

Meristem culture has also been used for rapid multiplication of individual red clover plants. This technique would primarily be useful for selected plant breeding methods such as double cross hybrids (See Chapter 12). Red clover plants can be regenerated rapidly by *in vitro* cultivation of isolated meristems (Churova, 1986). This research used surface sterilized apical meristems plated on MS basal salts with 0.6 μM IAA and 18 μM BA. Multiple shoots were produced from cultured meristems. These shoots then were rooted on MS with no growth regulators. The plants obtained were morphologically normal after transfer from *in vitro* to *in vivo* conditions. No evidence of somaclonal variation was observed.

Other research has used meristem culture to maintain and multiply useful genotypes of red clover cvs. Start, Kvarta, and Tatra, *T. repens, T. ambiguum*

Table 14.2

Comparison of organic composition of four media used for red clover tissue culture.

Compound	B5	L2	MS	EC6
	mg/l	mg/l	mg/l	g/l
Myo-inositol	100.0	250.0	100.0	
Nicotinic acid	1.0		0.5	
Pyroxidine-HCl	1.0	0.5	0.5	
Thiamine-HCl	10.0	2.0	0.1	
Glycine			2.0	
Yeast Extract				1.0

cv. Summit, *T. medium, T. sarosiense, T. rubens,* and interspecific hybrids of red clover with *T. medium.* The L2 basal salts supplemented with 2 μM adenine, 0.1 μM BA, and 0.01 μM NAA may be used for culturing all the above species and hybrids (Repkova, 1989a). B5 medium supplemented with BA and NAA has been used for red clover micropropagation by reducing basal salts to half-strength at the second subculture (Kir'yan et al. 1990). Young flower heads and bracts have been used as explant sources for rapid propagation of selected red clover genotypes (Cebrat et al., 1991). Histological studies showed that regeneration in this system occurred with omission of callus which is advantageous for genetic stability.

3. Callus induction and plant regeneration

In research leading to development of the L2 medium, this medium was shown to be broadly supportive of callus development from a number of red clover genotypes from five different cultivars (Phillips and Collins, 1979a). For callus induction these workers recommended picloram at 0.06 mg/l (0.25 μM) and BA at 0.1 mg/l (0.05 μM). Conversely, B5 basal salts supplemented with 11 μM NAA, 10 μM 2,4-D, and 10 μM kinetin have also been used successfully for red clover callus culture (Beach and Smith, 1979). Other combinations of auxins and cytokinins were tested, but were inferior.

Both of the above callus induction steps then were followed by a transfer to a second medium for shoot-bud development. Phillips and Collins recommended transfer to L2 with reduced auxin (picloram at 0.006 mg/l) and BA at 0.1 to 10 mg/l. The second step of the Beach and Smith system was B5 salts with an additional 10 mg/l thiamine (total of 20 mg/l), 10 μM NAA, and 15 μM adenine. Both groups recommended transfer of developing shoots to

Fig. 14. 1. Variable regeneration among red clover plants: Above - Highly regenerative genotypes, Below - nonregenerative genotypes.

a rooting medium after about four weeks on the shoot development medium. Rooting media of both groups were generally reduced (0.5×) basal salts and low growth regulator levels.

Initial explant sources for Phillips and Collins were variable, but the authors state that callus was induced from several different mature plant sources. Beach and Smith used hypocotyl and ovary (whole pistil) explants. More recently research has shown that petiole sections can also be used effectively as explant source for plant regeneration using the Beach and Smith protocol (Quesenberry and Smith, 1993).

Regeneration using either of the above systems was shown to be highly genotype dependent although selected genotypes from a number of cultivars showed limited regeneration. Other research in Canada and Europe has confirmed that regeneration was dependent on explant source and plant genotype (Kallak and Hilpus, 1992; MacLean and Nowak, 1989; Oelck and Schieder, 1983; Repkova, 1990; Solodkaya and Novoselova, 1988). As an example, L2 medium was superior for callus induction, but among 40 genotypes representing eight *Trifolium* spp. embryoids were formed only in two genotypes of 4× red clover and one of *T. pannonicum* acq., all using petioles as explant sources (Horak and Fojtik, 1984). As discussed in the next section, media and genotype manipulations which favor regeneration via somatic embryogenesis are desirable.

One solution to the genotype dependency problem is to breed for a population that has high regenerative capacity. This method was shown to be effective in alfalfa (Bingham et al., 1975) and resulted in the development and release of the Regen-S germplasm (Bingham, 1989). Callus morphology and plant regeneration in red clover were governed by highly significant

Fig. 14. 2. Stages in the regeneration of plants from red clover callus from early regenerants to young plants in peat ball.

additive genetic effects. Results suggested that the frequency of regenerable genotypes in a population should be a trait that could be increased by genetic selection (Keyes et al., 1980). Recurrent selection methods have also been applied successfully in red clover to increase regeneration frequency (Quesenberry and Smith, 1993). In a base population from the cultivar Arlington, eight plants which showed limited plant regeneration were selected out of 200 (4%) and intercrossed. Five cycles of selection (one cycle per year) were carried out. Percentage of regenerating plants in the base, recurrent phenotypic selection cycle 1 (RPS1), RPS2, RPS3, RPS4, and RPS5 was 4, 9, 16, 20, 41, 72, respectively (Fig. 14.1). Number of regenerating plants per petiole explant was greater than 200 for selected RPS5 individuals compared with less than 10 in base population individuals. Figure 14.2 shows the various stages in development of plants from red clover callus culture. Narrow-sense heritability estimates ranged from 40 to 50% showing that plant regeneration from callus tissue of red clover is a highly heritable trait.

An interesting corollary to genetic potential for regeneration is the report that regenerative genotypes of red clover were also more tolerant to cold stress in a controlled environment (Nowak et al., 1992). These workers identified genotypes from the cultivar Florex which varied in regenerative ability from callus and cell culture. Plantlets were hardened in culture tubes at 2 or 5 °C and 8 h photoperiod for four weeks and then subjected to -10 °C. Survival of regenerative genotypes was significantly higher than of the non-regenerative genotypes. This work suggests that *in vitro* selection for the regenerative trait could improve cold stress tolerance in red clover.

Recent research with white clover has resulted in the identification of a zone of plant tissue in seedlings which appears to produce direct plant regeneration relatively independent of plant genotype (White and Voisey, 1994). In this work cotyledons were dissected from 3 day old seedlings by bisecting the hypocotyl along the darker green band which is at the interface between the cotyledonary stalk, hypocotyl, and shoot apex. These cotyledon pairs were separated and the apical meristem was removed. Individual cotyledons were placed adaxial side up on MS salts medium. Superior plant regeneration was obtained with 1.0 mg/l BAP and 0.05 mg/l NAA as growth regulators. From 88 to 100% of genotypes from eight different cultivars of white clover formed shoots in this system. This protocol then was used to produce transgenic white clover (Voisey et al., 1994). The utility of this system in red clover needs to be investigated.

4. Somatic embryogenesis and somaclonal variation

A major concern with any regeneration system is maintenance of genetic integrity of the genotype which was placed into culture. Somaclonal variation induced while cells are in an undifferentiated state has been demonstrated in alfalfa and other legumes (Bingham and McCoy, 1986). This phenomenon is a concern in both meristem culture and cell and callus culture research. Reducing the time tissues remain in culture appears to reduce occurrence of somaclonal variation. Direct budding of meristems without callus production will generally not result in somaclonal variation (Phillips and Collins, 1979a; Cebrat et al., 1991).

Regeneration from callus via somatic embryogenesis has been suggested to reduce somaclonal variation. Phillips and Collins (1980) reported that cell suspension cultures maintained on SL2 liquid medium (SL2 is L2 with slightly reduced levels of most major and minor salts produced vigorous callus colonies when plated on agar solidified full-strength L2. Their research showed that optimal growth regulators for induction of somatic embryogenesis were 0.01 mg/l 2,4-D (0.045 μM) and 2.0 mg/l adenine (14.8 μM). With this medium 35 to 40% of plates showed developing buds. This medium was designated LSE (legume somatic embryo). For efficient shoot propagation, subculture of developing buds to L2 with 0.001 mg/l picloram (0.004 μM) and 0.2 mg/l BA (0.88 μM) was optimum. This medium was designated LSP (legume shoot promotion). Most buds in this system developed by somatic embryogenesis with emergence of a pair of cotyledons followed by a single unifoliolate leaf as occurs in normal red clover seed germination. Nevertheless, these authors reported that only 14 of 22 regenerated plants were normal diploids. Two were diploid with faciated stems and four were diploid with flower head and leaf morphology abnormalities and partial sterility. The remaining two were tetraploid suggesting spontaneous doubling in culture.

The NEWRC high frequency regeneration population of red clover (Smith and Quesenberry, 1995) developed by Quesenberry and Smith (1993) also regenerated primarily by somatic embryogenesis on the modified B5 protocol of Beach and Smith (1979). These workers reported no obvious morphological variation among over 50 plants regenerated from a single petiole explant. Later unpublished results with larger numbers of plants have shown only a very low frequency of abnormal regenerated plants. This finding strengthens the rationale for use of somatic embryogenesis systems in red clover where maintenance of genetic integrity is important.

Other research has also demonstrated direct somatic embryogenesis in red clover using various growth regulator and explant modifications. Immature embryos of *T. repens* and red clover have been cultured on EC6 basal medium with 0.025 to 0.05 mg/l BA and 1.0 g/l yeast extract to produce direct somatic embryos (Maheswaran and Williams, 1986, 1987). A one step system for somatic embryogenesis in red clover and *T. rubens* also was described (McGee et al., 1989). The ability of somaclones to regenerate in this work was shown to be related to translational patterns of two proteins isolated by two-dimensional gel electrophoresis. Explant source can affect ability of cultures to regenerate via somatic embryogenesis (Bhojwani et al., 1984). In their research only long term callus cultures from hypocotyls of red clover produced somatic embryos, however, somatic embryos were produced on callus formed directly on cultured anthers. Direct somatic embryogenesis of red clover cv. Start (2x) and cv. Kvarta (4x), *T. alpestre, T. ambiguum* cv. Summit, *T. medium, T. pannonicum, T. rubens,* and *T. vesiculosum* savi cv. Yuchi using immature zygotic embryos also has been reported (Repkova, 1989b). Culture was most successful in this work using torpedo stage embryos. Somatic embryogenesis was initiated on EC6 medium with 0.22 or 0.44 μM BA and 1.0 g/l yeast extract. A second transfer to LSE stimulated plantlet development.

Various researchers have proposed that somatic embryos could be dried and encapsulated in synthetic polymers to produce synthetic seed. Following this philosophy one superior single cross hybrid could be micropropagated and seed planted. Due to the extensive land area planted to red clover and relatively low value of the harvested crop, it is unlikely at present that such a procedure can be streamlined and developed on a cost effective basis. Nevertheless, other research applications of somatic embryos have been investigated. Workers in Japan (Sakai et al., 1991; Yamada et al., 1991) have reported successful vitrification and cryopreservation of red clover somatic embryos. This technique may be useful for long term storage of novel genotypes.

Fig. 14. 3. Somaclonal variant of red clover selected on a low P medium.

Nevertheless, concerns regarding somaclonal variations induced in some red clover tissue culture systems merit consideration. Several Russian workers have shown the occurrence of somaclonal variation (Stanis et al., 1983; Ginka, 1985). An example is a report where 59 of 98 regenerated plants from the diploid cv. VIK7 were diploid and the remainder were $2n = 4\times = 28$ (Mezentsev and Lyubavina, 1982). Some regenerates had altered flower structures with sterile pollen. Other work has shown that plants derived from hypocotyl and shoot tip explants of the cultivars Altasweed ($2\times$) and Norseman ($4\times$) through multiple shoot production via organogenesis had stable isoenzyme banding patterns for five enzyme systems studied and stable nodule leghemaglobin profiles. However, variations were detected for chromosome number, leaflet length:width ratio, and leaflet number (Wang and Hall, 1988).

A low P medium has been used to select for somaclonal variability in red clover (Phillips and Collins, 1981). When regenerated somaclonal plants were evaluated in the greenhouse, many had serious fertility problems (Bagley and Taylor, 1987). Most regenerated plants did not flower or did so only sporadically, although all had normal $2n = 14$ chromosomes (Fig. 14.3). Meiotic analysis of somaclone C3-N showed 13% of the pollen mother cells at the tetrad stage with cell numbers different from the normal four. Pollen stainability among four somaclones varied from a near normal 97% to as low as 58%.

5. Protoplast culture

With development of improved media for tissue and cell culture of red clover, other research focused on development of protocols for isolation and culture of red clover protoplasts. Initial technique research involved red clover and the related species *T. rubens* (Grosser and Collins, 1983, 1984) . These workers used an enzyme solution modified from Kao and Michayluk (1975). Aseptically grown leaves were feathered with a scalpel, infiltrated with the enzyme solution under vacuum, and incubated 4 to 6 h on a rotary shaker in the dark. This solution then was filtered through a 45 *u*m mesh screen to remove undigested cell clumps and debris and then centrifuged for 6 min at 100 × g. The pellet was resuspended in the 8P protoplast culture medium of Kao and Michayluk. The resuspension-centrifugation was repeated two times to remove all enzyme. Protoplasts also were isolated from suspension cultures maintained on SL2 which had been established from seedling hypocotyls. A similar isolation procedure as with leaves was followed. Protoplasts initially were cultured in liquid 8P medium, but at 7 to 14 days were supplemented with macro- and micronutrients of SL2. Finally, cell colonies were plated on agar solidified L2 and grown for one month. Large calli then were transferred to LSE (Phillips and Collins, 1980) for production of somatic embryos and shoots. Shoots were propagated on LSP.

Leaf-derived protoplasts were generally smaller and more uniform in size than suspension culture derived-protoplasts. Chloroplasts were conspicuous in the leaf-derived protoplasts. Cell division of leaf-derived protoplasts occurred at 6 days after isolation. Suspension culture-derived protoplasts were variable in size and began cell wall synthesis almost immediately after isolation. Only mature plants of *T. rubens* were recovered using this system. Plants regenerated from leaf-derived protoplasts were 2n = 2× = 16 whereas those from suspension culture-derived protoplasts were 2n = 4× = 32. The authors speculated that the suspension culture had become 4× prior to isolation of protoplasts. Using the chloroplast differences between suspension culture-derived protoplasts and leaf-derived protoplast, heterokaryons were easily identified, however, no intact plants of *T. rubens* × red clover were produced (Grosser and Collins, 1983).

Workers in Finland and the Ukraine have reported isolation and culture of red clover and *T. hybridum* protoplasts (Honkanen et al., 1986, 1987; Kuchuk, 1989). Similar to Grosser and Collins, no mature plants of either red clover or *T. hybridum* were obtained from isolated protoplasts (Honkanen et al., 1986, 1987) . They did obtain dividing cells and callus formation. These workers reported heterokaryon formation following protoplast fusion of these two species, however, no mature hybrid plants were produced. Son et al. (1989) reported on factors affecting the isolation of red clover mesophyll protoplasts and presented a modified enzyme mixture and protocol for protoplast isolation. As with regeneration from suspension or callus culture, regeneration from protoplasts also appeared to be genotype dependent (Myers, et al., 1989). In this research, one genotype was identified which had regeneration potential similar to *T. rubens*, a species which had high regeneration potential from protoplast culture (Parrott and Collins, 1982). Regeneration of this genotype was achieved from both leaf and suspension culture-derived red clover protoplasts.

6. *In vitro* selection

A primary reason for the broad scope of interest in tissue culture of red clover has been the potential for application of cellular selection techniques. A second potential application is somatic cell fusion, particularly of red clover with related perennial species. Cellular selection research in red clover has focused in three major areas: 1) disease resistance selection; 2) herbicide tolerance selection; and 3) low inorganic mineral uptake or tolerance.

Various European and Canadian workers have reported cellular selection research using culture filtrates of *Fusarium oxysporum*, *F. roseum*, and *Sclerotinia trifoliorum* added to red clover cell culture media. All these researchers have primarily relied on the production of somaclonal variation induced in culture as a source of variability on which to select. Some have used a wide variety of cultivars and individuals within cultivars as sources for cell cultures. *In vitro* selection for growth in the presence of *S. trifoliorum* filtrate increased the percentage of resistant plants about 65% in three different cell lines derived from three different cultivars (Solodkaya, 1987; Solodkaya and Novoselova, 1989). *S. trifoliorum* resistant plants were also selected from the cultivar Japioi Poly using *in vitro* selection against culture filtrate (Mezentsev et al., 1984). Regenerated plants were confirmed as resistant in field trials under artificial infection. The successful use of *in vitro* selection methods to select red clover plants resistant to *F. roseum* and *F. oxysporum* has been reported (Constabel, 1989; Atkociunaite et al., 1991).

In vitro selection for tolerance to 2,4-D has been conducted in red clover (Taylor et al., 1989a). In this research, *in vivo* and *in vitro* responses were shown to be highly correlated ($r = 0.77$). After two cycles of selection on L2 medium supplemented with 0.18 mM 2,4-D, 16 tolerant callus lines were identified based on visual growth. Amount of 2,4-D present in these callus clones was evaluated using an oat coleoptile elongation bioassay. Two of the more tolerant lines had 61 and 83% less 2,4-D in their tissue than the susceptible control. However, no mature plants could be regenerated from this selected callus.

These workers also used callus culture to measure the relative susceptibility of red clover to 2,4-D vs. 2,4-dichlorophenol (2,4-DCP), a metabolite of 2,4-D. They demonstrated that red clover callus was significantly more tolerant to 2,4-DCP than to 2,4-D. This suggests the possibility of improving red clover resistance to 2,4-D by incorporating a gene into red clover via genetic transformation which could metabolize 2,4-D to 2,4-DCP. Such a gene is known in microorganisms (Taylor et al., 1989b). *In vitro* methods also were evaluated as a technique to identify tolerance to the herbicide imaquazin, but the authors concluded that for this herbicide, *in vivo* methods were superior. (Taylor et al., 1986).

The ability of red clover cells to grow on reduced or elevated levels of inorganic minerals has been suggested as an *in vitro* selection method. Red clover somaclonal variants have been selected under low P concentration in suspension culture. Callus growth on low P media was correlated to whole plant P and N concentration (Phillips and Collins, 1981). When regenerated plants from these selected somaclones were grown in low P soils, no difference in yield was found compared to unselected plants (Bagley and Taylor, 1987).

7. Ovule and embryo culture

Most of the applications of embryo culture in red clover have been discussed in chapter 2 as related to interspecific hybrids. Although the early work of Evans (1962) was not successful, evidence continued to accumulate that endosperm breakdown was the primary cause for failure of a number of hybrids. The first successful application of embryo rescue techniques to red clover was the interspecific hybrid of *T. sarosiense* × red clover (Phillips, et al., 1982). In this work two new media were developed for culture of immature interspecific hybrid embryos. Optimal medium for initial culture of embryos was determined to be L2 basal salts supplemented with 12.5% sucrose, 15 *u*M adenine, and 25 μM picloram. This initial culture medium was used for 8 to 14 days, after which embryos would turn yellow or white and cease to grow. Thus, prior to degeneration on the LIH medium embryos were transferred to a second medium, designated LSP2, containing L2 basal medium with the standard 2.5% sucrose, 0.66 μM BA, and 4 μM picloram. Only about half of the embryos produced shoots directly on the LSP2 medium, the remainder were transferred to LSE (Phillips and Collins, 1980) in attempts to obtain additional shoots. Only one additional F_1 was rescued by this method. Of 51 F_1 crossing combinations originally attempted, 5 unique F_1 families were grown to maturity. The best compromise between maximum number of viable embryos and maximum degree of development appeared to be 14 to 19 days after pollination. Predominantly heart and early torpedo stage embryos were obtained at this stage. Other *Trifolium* interspecific hybrids were best rescued at 12–14 days after pollination (Sawai et al., 1990).

The media developed in the above research have been used to produce interspecific hybrids of *T. medium* × red clover (Merker, 1984; Sawai, et al., 1990). Media composition studies with this hybrid showed that L2 supplemented with 0.02 mg/l 2,4-D, 0.005 mg/l para- chlorophenoxyacetic acid (PCA), and 0.4 mg/l adenine was optimum (Sawai et al., 1990), whereas Merker (1984) was successful using picloram and adenine as described by Phillips et al. (1982). The protocols of Phillips, et al. (1982) were also used to rescue the *T. alpestre* × red clover interspecific hybrid (Merker, 1988; Phillips et al., 1992). Recently, hybrids of white clover form *hollandicum* × red clover cv. Start (diploid) and cv. Tatra (tetraploid) have been reported through embryo rescue (Repkova et al., 1991). These workers rescued heart shaped embryos onto LIH basal medium with transfer to medium with gradually decreasing levels of sucrose. Most embryos were rescued from crosses using white clover as the female, where 104 plants were obtained. Isoenzyme patterns of 24 hybrid plants were intermediate between the parental isoenzyme patterns.

8. Genetic transformation

Most methods of plant genetic transformation require that individual undifferentiated cells be exposed to the transforming agent, that foreign DNA be incorporated into the host chromosome, and that individual cells then be regenerated to whole plants. Thus callus, suspension, or protoplast cultures which will readily regenerate are a requirement for application of these molecular techniques to red clover. Recently, red clover carrying the NPTII gene for kanamycin resistance has been reported (Quesenberry et al., 1992; Strukcinskaite et al., 1991; Ivashuta et al., 1991). Little detail is available for the latter two works, but apparently both groups used kanamycin resistance to select for transformed plants. The use of *Agrobacterium tumefaciens* strain A81 carrying a binary vector pPCV002 with kanamycin resistance was reported (Strukcinskaite et al., 1991). Both groups reported use of 200 mg/l claforan to eliminate the *Agrobacterium* after co-cultivation and 50 mg/l kanamycin as the selective agent to allow growth of only transformed plants.

In Florida, petiole explants of red clover were tolerant of up to 340 mg/l of carbenicillin but callus growth was inhibited completely by greater than 25 mg/l kanamycin (Quesenberry et al., 1992). To insure survival of only transformed plants these workers used 50 mg/l of kanamycin in the selective media similar to the above two reports. This allowed use of the former antibiotic to kill *Agrobacterium tumefaciens* after transformation, and selection of transformed cells by growth on kanamycin containing medium. Using four red clover genotypes and three *Agrobacterium* strains, they showed significant genotype, strain, and interaction effects for success of transformation. No transformants were obtained using an *Agrobacterium* strain which contained a

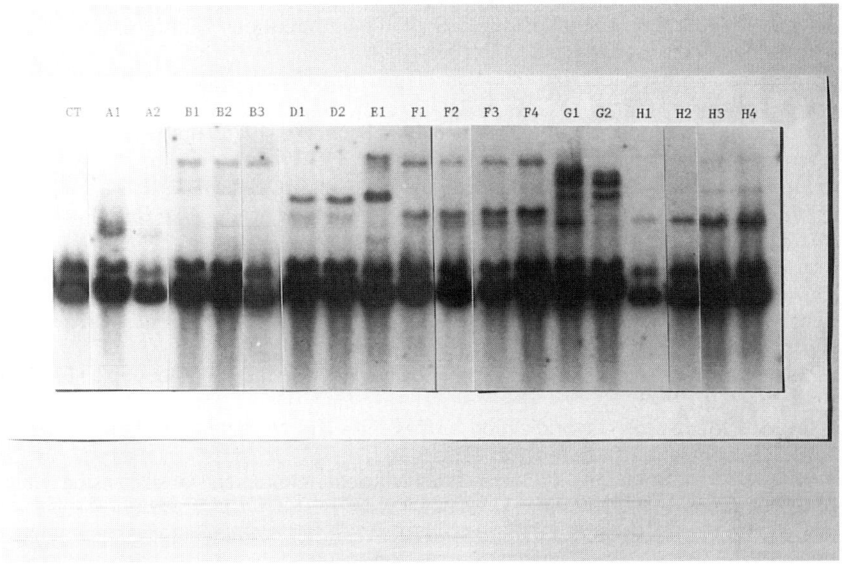

Fig. 14. 4. Southern blot of transgenic red clover plants. CT = Normal non-transformed, Letters refer to different callus clones, and numbers refer to regenerates within callus clones.

cointegrated plasmid, while both strains containing the binary plasmid yielded transformants. Differences were detected among genotypes in percentage of petiole explants producing transformed callus and one genotype yielded no transformed callus. Up to 50% of the petioles of the best clover genotype × *Agrobacterium* strain combinations produced transformed callus. All transformed plants were uniformly positive for the GUS reaction. Presence of the NPTII gene in transformed plants was confirmed by Southern blots. All plants which came from a given piece of callus had the same banding pattern on Southern blots, suggesting that all regenerated plants from a given piece of callus carry the same gene insertion (Fig. 14.4). Recent unpublished results show that F_1 progeny of crosses of transformed × normal segregated 1:1 as expected.

References

Atkociunaite, V., B. Grigaliunaite and R. Janaudyte. 1991. Selection of red clover cell lines and regeneration of plants resistant to the toxins of *Fusarium oxysporum* Schl. *Ekxperimentine Biologija.* No. 1:20.

Bagley, P.C. and N.L. Taylor. 1987. Evaluation of phosphorus efficiency in somaclones of red clover. *Iowa State J. Res.* 61:459–480.

Beach, K.H. and R.R. Smith. 1979. Plant regeneration from callus of red and crimson clover. *Plant Sci. Lett.* 16:231–237.

Bhojwani, S.S., K. Mullins and D. Cohen. 1984. Intra-varietal variation for *in vitro* plant regeneration in the genus *Trifolium*. *Euphytica.* 33:915–921.

Bingham. E.T. 1989. Registration of Regen-S alfalfa germplasm useful in tissue culture and transformation research. *Crop Sci.* 29:1095–1096.

Bingham, E.T. and T.J. McCoy. 1986. Somaclonal variation in alfalfa. *Plant Breed. Rev.* 4:123–151.

Bingham, E.T., L.V. Hurley, D.M. Kaatz and J.W. Saunders. 1975. Breeding alfalfa which regenerates from callus tissue in culture. *Crop Sci.* 15:719–721.

Cebrat, J., H. Kruczkowska, W. Miszke, H. Pawlowska and B. Skucinska. 1991. Micropropagation of red clover (*Trifolium pratense* L.) from flower heads. *Acta Biol. Cracov. Ser. Bot.* 32:235–242.

Churova, K. 1986. Meristem culture and micropropagation of red clover (In Slovakian). *Rostlinna-Vyroba.* 32:411–415.

Constabel, E.C. 1989. *In vitro* selection of red clover for resistance to *Fusarium roseum* L. and evaluation of regenerated plants. *Forage Notes.* 34:78.

Dale, P.J. and V.A. Cheyne. 1993. The elimination of clover diseases by shoot tip culture. *Ann. Appl. Biol.* 123:25–32.

Evans, A.M. 1962. Species hybridization in *Trifolium* II. investigating the pre-fertilization barriers to compatibility. *Euphytica.* 11:256–262.

Gamborg, O.L., R.A. Miller and K. Ojima. 1968. Nutrient requirements of suspension cultures of soybean root cells. *Exp Cell Res.* 50:148–151.

Gamborg, O.L., T. Murashige, T.A. Thorpe and I.K. Vasil. 1976. Plant tissue culture media. *In Vitro.* 12:473–478.

Ginka, O.V. 1985. Morphological features of red clover regenerants produced by means of cell and tissue culture. (In Russian) Netraditsionnye metody selektsii zernovykh kul'tur i kormovykh trav v Severo-Zapadnoi zone RSFSR. pp. 129–133.

Grosser, J.W. and G.B. Collins. 1983. Isolation, culture, and fusion of *Trifolium pratense* and *T. rubens* protoplasts. p. 65. In *Agronomy Abstracts.* ASA-CSSA-SSSA Madison, WI.

Grosser, J.W. and G.B. Collins. 1984. Isolation and culture of *Trifolium rubens* protoplasts with whole plant regeneration. *Plant Sci. Lett.* 37:165–170.

Honkanen, J., P. Ryoppy and P.M.A. Tigerstedt. 1986. Protoplast culture and fusion of red and alsike clover. In W. Horn, C.J. Jensen, W. Odenbach and O. Schieder (eds.), *Proc. Intl. Symp. Eucarpia.* Berlin 8–13 Sep. 1985. Walter de Gruyter, Berlin.

Honkanen, J., P. Ryoppy and P.M.A. Tigerstedt. 1987. Protoplast fusions with clovers. *Nordisk Jordbrugsforskning.* 69:213.

Horak, J. and A. Fojtik. 1984. Callus formation and plant regeneration in tissue culture of genus *Trifolium.* In F.J. Novak, L. Havel and J. Dolezel (eds.), *Plant Tissue and Cell Culture Application to Crop Improvement.* Prague, Czechoslovakia.

Ivashuta, S.I., V.V. Mazin and L.S. Solodkaya. 1991. Genetic transformation of red clover bud culture. (In Russian) 1st Vsesoyuznyi simposium "Novye metody biotekhnologii rastenii," Pushchino, 20–22 November, 1991. Tezisy Dokladov. 18–19:126–127.

Kallak, H. and I. Hilpus. 1992. On genetic aspects of *in vitro* morphogenesis in plants. *Ekxperimentine Biologija.* 3–4:39–40.

Kao, K.N. and M.R. Michayluk. 1975. Nutritional requirements for growth of *Vicia hajastana* cells and protoplasts at a very low population density in liquid media. *Planta.* 126:105–110.

Keyes, G.J., G.B. Collins and N.L. Taylor. 1980. Genetic variation in tissue cultures of red clover. *Theor. Appl. Genet.* 58:265–271.

Kir'yan, I.G., V.V. Mazin and A.S. Shain. 1990. Methodology for clonal micropropagation of red clover (*Trifolium pratense* L.) in culturing lateral buds. (In Russian). *Sel'skhozyaistvennaya-Biologiya.* 5:94–97.

Kuchuk, N.V. 1989. Extraction and culture of protoplasts of five species of the Leguminosae. *Fiziologiya Rastenii.* 36:821–824.

MacLean, N.L. and J. Nowak. 1989. Plant regeneration from hypocotyl and petiole callus of *Trifolium pratense* L. *Plant Cell Rept.* 8:395–398.

Maheswaran, G. and E.G. Williams. 1984. Direct somatic embryoid formation on immature embryos of *Trifolium repens, T. pratense* and *Medicago sativa*, and rapid clonal propagation of *T. repens*. *Ann. Bot.* 54:201–211.

Maheswaran, G. and E.G. Williams. 1986. Clonal propagation of *T. pratense, T. resupinatum,* and *T. subterraneum* by direct somatic embryogenesis on cultured immature embryos. *Plant Cell Rep.* 5:165–168.

Maheswaran, G. and E.G. Williams. 1987. Uniformity of plants regenerated by direct somatic embryogenesis from zygotic embryos of *Trifolium repens*. *Ann. Bot.* 59:93–97.

McGee, J.D., E.G. Williams, G.B. Collins and D.F. Hildebrand. 1989. Somatic embryogenesis in *Trifolium*: protein profiles associated with high- and low-frequency regeneration. *J. Plant Physiol.* 135:306–312.

Merker, A. 1984. Hybrids between *Trifolium medium* and *Trifolium pratense*. *Hereditas.* 101:267–268.

Merker, A. 1988. Amphidiploids between *Trifolium alpestre* and *Trifolium pratense*. *Hereditas.* 108:267.

Mezentsev, A.V. and L.A. Lyubavina. 1982. Regeneration of plants from cell suspension cultures of *Trifolium pratense*. (In Russian). *Fisiologiya Rastenii.* 29:1006–1009.

Mezentsev, A.V., N.A. Golyshkina and L.A. Solodkaya. 1984. Producing initial breeding material of clover and lucerne using cell and tissue culture methods. (In Russian). *Sbornik Nauchnykh Trudov Vsesoyuznogo Nauchno issledovatel'skogo Instituta Kormov.* 31:57-62.

Murashige, T. and F. Skoog. 1962. A revised medium for rapid growth and bioassays with tobacco tissue culture. *Physiol. Plant.* 15:473–479.

Myers, J.R., J.W. Grosser, N.L. Taylor and G.B. Collins. 1989. Genotype-dependent whole plant regeneration from protoplasts of red clover (*Trifolium Pratense* L.). *Plant Cell Tissue Organ Cult.* 19:113–127.

Nowak, J., S.L. Matheson, N.L. MacLean and P. Havard. 1992. Regenerative trait and cold hardiness in highly productive cultivars of alfalfa and red clover. *Euphytica.* 59:189–196.

Oelck, M.M. and O. Schieder. 1983. Genotypic differences in some legume species affecting redifferentiation ability from callus to plants. *Plant Breed.* 91:312–321.

Parrott, W.A. and G.B. Collins. 1982. Callus and shoot tip culture of eight *Trifolium* spp. *in vitro* with regeneration via somatic embryogenesis of *Trifolium rubens*. *Plant Sci. Lett.* 28:189-194.

Phillips, G.C. and G.B. Collins. 1979a. *In vitro* tissue culture of selected legumes and plant regeneration from callus cultures of red clover. *Crop Sci.* 19:59–64.

Phillips, G.C. and G.B. Collins. 1979b. Virus symptom-free plants of red clover using meristem culture. *Crop Sci.* 19:213–216.

Phillips, G.C. and G.B. Collins. 1980. Somatic embryogenesis from cell suspension cultures of red clover. *Crop Sci.* 20:323–326.

Phillips, G.C. and G.B. Collins. 1981. Growth and selection of red clover *Trifolium pratense* L. cells on low levels of phosphate. p. 187. In *Agronomy Abstracts.* ASA-CSSA-SSSA Madison, WI.

Phillips, G.C., G.B. Collins and N.L. Taylor. 1982. Interspecific hybridization of red clover (*Trifolium pratense* L.) with *T. sarosiense* Hazsl. using *in vitro* embryo rescue. *Theor. Appl. Genet.* 62:17–24.

Phillips, G.C., J.W. Grosser, S. Berger, N.L. Taylor and G.B. Collins. 1992. Interspecific hybridization between red clover and *Trifolium alpestre* using *in vitro* embryo rescue. *Crop Sci.* 32:1113–1115.

Quesenberry, K.H. and R.R. Smith. 1993. Recurrent selection for plant regeneration from red clover tissue culture. *Crop Sci.* 33:585–589.

Quesenberry, K.H., D.S. Wofford, P.A. Krottje and R.L. Smith. 1992. Production of transgenic red clover. pp. 5–10. In D.S. Wofford and K.H. Quesenberry (eds.), *Proc. Twelfth Trifolium Conf.*, Gainesville, FL. 25–27 March 1992.

Repkova, J. 1989a. Regeneration of clovers from meristems for interspecific hybridization. (In Czech). *Sb. Ved. Pr. Vyzk. Slechtitelskeho Ustavu Picninarskeho Troubsku Brna.* 11:5–11.

Repkova, J. 1989b. Direct somatic embryogenesis and plant regeneration from cultured embryos in the genus *Trifolium* (In Czech). *Sbornik UVTIZ Genetika a Slechteni.* 25:173–180.

Repkova, J. 1990. Callus formation and regeneration in cultures of explants from various red clover genotypes (In Czech). *Sbornik UVTIZ Genetika a Slechteni.* 26:71–76.

Repkova, J., B. Nedbalkova and J. Holub. 1991. Regeneration of plants from zygotic embryos after interspecific hybridization within the genus *Trifolium* and electrophoretic evaluation of hybrids. *OSEVA Research Institute for Fodder Plants.* 12:7–14. (From Plant Breeding Abstracts 1993).

Rupert, E.A. and G.B. Collins. 1985. Tissue culture. In N.L. Taylor (ed.), Clover Science and Technology. ASA- CSSA-SSSA 25:405–416 Madison, WI.

Sakai, A., S. Kobayashi and I. Oiyama. 1991. Survival by vitrification of nucellar cells of navel orange (*Citrus sinensis* var brasiliensis Tanaka) cooled to -196 C. *J. Plant Physiol.* 137:465-470.

Sawai, A., S. Ueda, M. Gau and K. Uchiyama. 1990. Interspecific hybrids of *Trifolium medium* L. × 4× *Trifolium pratense* L. obtained through embryo culture. *J. Japan. Soc. Grassl. Sci.* 33:157–162.

Servitova, J. and R. Pokorny. 1987. Curing red clover of virus diseases using the method of meristem culture. (In Czech). *Sb. Ved. Pr. Vyzk. Slechtitelskeho Ustavu Picninarskeho Troubsku Brna.* 10:143–148.

Smith, R.R. and K.H. Quesenberry, 1995. Registration of NEWRC red clover germplasm. *Crop Sci.* 35:295.

Solodkaya, L.A. 1987. *In vitro* selection and evaluation of somaclonal variants of red clover. (In Russian) *Gametnaya i zigotnaya selektsiya rastenii. Respublikanskaya konferentsiya,* 23 June 1987. pp 187–188. Kishinev, Moldavian SSR.

Solodkaya, L.A. and E.Y. Novoselova. 1988. Physiological and genotypic features of embryogenesis and morphogenesis in suspension cultures of red clover. (In Russian). *Biologiya kul'tiviruemykh i biotekhnologyia.* 1:119–120.

Solodkaya, L.A. and E.Y. Novoselova. 1989. Methods of producing red clover with high resistance to rot. (In Russian). *Sbornik Nauchnykh Trudov Vsesoyuznyi Nauchno Issledovatel'skii Institut Kormov.* 42:57–65.

Son, D.Y., B.W. Yun and J.K. Jo. 1989. Some factors affecting the isolation of mesophyll protoplasts from red clover (*Trifolium pratense* L.). *J. Korean Soc. Grassl. Sci.* 9:119–123.

Stanis, V.A., A.K. Slesaravichyus and V.A. Kaleda. 1983. Regeneration of *Lolium perenne* and *Trifolium pratense* plants by culturing tissue of different origin. (In Russian). *Sel'skokhozyaistvennaya Biologiya.* 5:29–31.

Strukcinskaite, J., I. Meskiene and V. Atkociunaite. 1991. Obtaining transgenic red clover using *Agrobacterium* binary vector. Eksperimentine Biologija No. 1, 42; In *Proc. 6th Congress of Lithuaniana Geneticists and Breeders.*

Taylor, S.G., J.A. Dusky and K.H. Quesenberry. 1986. *In vivo* and *in vitro* methods of identifying tolerance of red clover to imazaquin herbicide. *Soil Crop Sci. Soc. Fl. Proc.* 46:56–59.

Taylor, S.G., D.G. Shilling and K.H. Quesenberry. 1989a. *In vitro* selection for 2,4-D tolerance in red clover. *Theor. Appl. Genet.* 78:265–270.

Taylor, S.G., D.G. Shilling and K.H. Quesenberry. 1989b. Phytotoxicity of 2,4-D and 2,4-dichlorophenol to red clover (*Trifolium pratense* L.). *Weed Sci.* 37:825–829.

Voisey, C.R., D.W.R. White, B. Dudas, R.D. Appleby, P.M. Ealling and A.G. Scott. 1994. *Agrobacterium*-mediated transformation of white clover using direct shoot organogenesis. *Plant Cell Rept.* 13:309–314.

Wang, H. and F.B. Hall. 1988. *In vitro* culture and the incidence of somaclonal variation in regenerated plants of *Trifolium pratense* L. *Plant Sci.* 55:159–167.

White, D.W.R. and C. Voisey. 1994. Prolific direct plant regeneration from cotyledons of white clover. *Plant Cell Rept.* 13:303–308.

Yamada, T., A. Sakai, T. Matsumura and S. Higuchi. 1991. Cryopreservation of apical meristems of white clover (*Trifolium repens* L.) by vitrification. *Plant Sci.* 78:81–87.

15. Seed Production

Seed of red clover is produced in many temperate regions of the world. Seed production is often a by-product of forage production. Most countries desire to be self sufficient in producing seeds of the crop, but may be limited by excessive rainfall during the growing season. For this reason, some specialized areas have been developed such as the Pacific northwest in the U.S. (Fig. 15.1), the south island of New Zealand, and the southern latitudes of Europe.

Seed production of clovers in general has been reviewed by Rinker and Rampton (1985), and no attempt will be made here to duplicate that review. Instead, the present review will concentrate only on the red clover literature since about 1985.

1. Cultural practices

1.1. *Sowing the crop*

Seedbed preparation for red clover seed production is about the same as for forage production. (See Chapter 5 for more details on establishment.) Stands may be established in spring or fall, depending upon climatic conditions. For example, in northeastern U.S., seeds are usually sown in the spring, but in the south and Pacific northwest, they may be sown in either spring or fall. In Poland, the usual recommendation is to undersow red clover in spring barley in early spring, or to sow alone in early spring, mid-August, or late August (Wilczek, 1986). The optimum seeding rate in Poland is the same for seed and forage, 8 to 10 kg/ha (Bawolski and Scibior, 1985). In eastern U.S., red clover is sown 10 to 15 kg/ha (Rincker and Rampton, 1985). Low seeding rates may be used in eastern and western U.S. areas if seed is limited. In France, the culture of red clover for seed is different from that for forage. If sown alone, the optimum seeding rate is 3 to 5 kg/ha for diploids, and 5 to 8 kg/ha for tetraploids. If undersown in a small grain, the respective rates are 6 to 7 and about 8 kg/ha (C. Mousset-Declas, personal communication).

Fig. 15. 1. Seed production of Kenland red clover near Madras, Oregon (USA).

1.2. *Management of the crop for optimum seed yields*

The first, second, and subsequent years growth may be allowed to produce seed. In eastern U.S., first-year stands will produce seed but yields may be lower than in the second; thus, first year crops generally are used for forage rather than seed. In specialized seed producing areas, the first growth may be saved for seed, particularly if fall-sown. Nevertheless, some spring seedings also may be saved for seed. In eastern Canada, yields of up to 580 kg/ha were obtained from seedling stands of five cultivars on two soil types (Belzile, 1994).

In certain areas, the first growth of red clover in a season is removed and only the second crop is saved for seed. For example, in the southern part of the U.S. clover belt, the first crop sets few seeds because of attacks by the lesser clover leaf weevil (*Hypera nigrirostris* F.) (See Chapter 9). Likewise, in Canada, a prior vegetative cutting is essential for a good seed yield. The optimum timing of this cutting may vary from bud stage to 50% bloom (Belzile, 1990). In Czechoslovakia, seed yields generally are highest from stands in which the first crop is cut at bud formation as contrasted to later dates (Binderva, 1985). In Kentucky, as a result of several studies, the optimum stage of bloom for removal of the first crop has been determined to be about 10 to 15 days after the first bloom appears in the field (Taylor et al., 1972). On the other hand, in western U.S., the later the first growth is mowed, the more the seed yields are reduced (Rincker and Rampton, 1985). Consequently, mowing of first growth is not recommended for that location. A practice followed in western U.S. is to allow the first and second growth seed crops to accumulate *in situ*, and to harvest seed of both growths at one time. Low rainfall and sub-irrigation preserves the seed in good condition. It appears that the optimum date for removal of the first growth, if any, must be determined for each particular situation and location.

The optimum stage of maturity to harvest red clover for seed is generally estimated for each field by examination of heads. In Ireland, the optimum time to harvest for seed production is about 3 or 4 weeks after the end of the period of rapid inflorescence production. This generally coincides with the time when only a small proportion of unripe inflorescences remain (Puri and Laidlaw, 1984). In dry climates when seed is produced with sub-irrigation, the harvest date may be delayed without danger of shattering.

2. Pest control

Control of pests is discussed primarily in Chapter 6, Fungal Diseases and Chapter 9, Insect Pests. It is often more economical to control pests for seed than for forage production.

3. Weed control

Weeds in red clover often can be controlled by mowing if it is not detrimental to stands, forage, and seed yields. Mowing may be recommended if it coincides with the optimum time of cutting red clover for optimum seed or forage yields. Mowing annual broadleaf weeds below their growing points usually will provide adequate control. If this is not possible, chemical controls may be necessary and economically feasible. In Russia, the most suitable herbicides for red, white, and alsike clover seed production are 2M-4Kh (MCPA), 2M-4KhM (MCPB), Basagran (bentazone), and chlor-1FK (chloropham) (Bydrina and Voskresenskaya, 1980). In Norway, MCPA plus bentazone has been found to be the best mixture to control weeds in the year of sowing, although this treatment was not specifically for seed production (Skuterud, 1982). Benefin or Benefin plus 2,4-DB was recommended for use in Bulgaria for control of weeds of red clover in the year of sowing, so that no control would be needed in the second year. These treatments surpassed the seed yield of controls by 9.6 to 38.5% (Benkov and Dimitrova, 1983). In Czechoslovakia, monocot and dicot weeds were controlled by Alirox 80 EC (EPTC + antidote + AD-67), Eradicane 6E (EPTC + antidote + R-25788), Fusilade super (Fluazifop-butyl) + alfalon 50WP (linuron). Synfloran 24 EC + Alfalon 50 WP and Sys 67 Buratol + labuctril 25 both caused yield reductions (Bojas, 1986). In France, the most common treatments are basagran (bentazone) + tropotone (MCPB) or Lentagran (pyridate) to control dicotyledonous weeds, and legurame (carbetamide) or specific foliar graminicide (fop-din) against grasses (C. Mousset-Declas, personal communication).

4. Pollination

Because red clover is cross pollinated, bees are needed for seed production. Honey bees (*Apis mellifera* L.) frequently are used for this purpose because the more effective bumble bees (*Bombus* spp.) are often in short supply. If no other sources of pollen are in the area, or if honey bees are caged, they will effectively pollinate red clover. In this case, five hives of honey bees per hectare in close proximity to the seed field are sufficient for adequate seed yields (Rincker and Rampton, 1985).

In Poland, up to 20 species of bumble bees have been recorded as pollinating red clover, *Bombus terrestris* and *B. lucorum* being the most common. As few as 1950 to 2450 bees/ha gave satisfactory seed yields (Ruszkowski et al., 1989). In south island of New Zealand, the major pollinators were *B. ruderatus* and *B. hortorum* (Macfarlane and Griffin, 1985). For the tetraploid Grassland Pawera cultivar, a single bumble bee pollinated enough flowers to produce 0.3 to 0.6 kg/ha of seed under reasonable growing conditions. For maximum seed set, the best flowering crops need 4000 bumble bees/ha, and average crops need half this number. Prevailing temperatures allowed 10 hours of flight per day by bumble bees whereas honey bees flew only 6.5 hours per day (Macfarlane et al., 1991).

Leafcutter bees, *Megachile rotundata* (F.), are sometimes used as pollinators in western U.S. and Canada, but have not been used much in the eastern states. This bee nests in boards in which holes have been drilled to form tunnels. Leafcutter bees as pollinators for red clover in the Peace River region of northern Alberta, Canada during the period 1983 to 1987 produced an average seed yield of 410 kg/ha compared to 291 kg/ha where the leafcutter bee was not used (Fairey et al., 1989). Another bee sometimes used in western U.S. is the alkali bee (*Nowia* spp.) which nests in soil. However, these bees are difficult to move and manage.

In addition to having adequate numbers of pollinators for high seed yields, it is also important to have a large number of inflorescences per unit area. In most fields, the major component of seed yield is number of flower heads rather than number of seeds per head, or weight per seed. In eastern U.S., at least 225 heads per m^2 are necessary for a seed yield of approximately 225 to 280 kg/ha (Taylor et al., 1966). In Washington (USA), up to 1323 heads/m^2 and a seed yield of 927 kg/ha have been recorded (Rincker and Rampton, 1985). In a recent study in Oregon, with large numbers of heads/m^2 (698 to 944 in 1991), lower seed yields as a result of water stress, were primarily associated with decreased floral fertility and reduced numbers of flowers per unit area (Oliva et al., 1994b). The average seed yield of diploids and tetraploids in France (1992 to 1994) was 470 kg/ha but numbers of seedheads are not available (C. Mousset-Declas, personal communication).

5. Means of increasing seed production

5.1. *Irrigation*

Supplemental water is needed in many seed producing areas to insure adequate yields. It may be applied in furrow, by flooding, or by sprinklers. Up to 140 cm of irrigation water may be applied during the growing season in the drier areas of eastern Oregon (Rincker and Rampton, 1985).

Recent investigations have shown benefit from supplemental water even in the more humid Williamette valley of Oregon. The use of a crop water stress index (CWSI) has been proven valuable for timing of irrigation (Olivia et al., 1994a). One irrigation filling the active soil profile during peak flowering in the first year was sufficient to provide high seed yields, and to reduce losses from root diseases in the second year (Oliva et al., 1994b).

5.2. *Growth hormones*

Alar (daminozide) is the chemical that has been used most often to increase seed production (Table 15.1). Most investigators report modest to high yield increases resulting primarily from increased head number. Application period was usually at the flower bud stage. The economics of the use of alar was not addressed.

Other chemicals investigated include the growth retardants CCC (chlormequal) and composan (ethephone) applied after the last cut in September. These chemicals decreased growth of plants in autumn and increased frost resistance, chlorophyll content, photosynthesis, and fresh herbage and seed yields (Trofimova et al., 1987). Fertilizer trace elements applied to red clover to increase nectar production had little or no effect according to Jakesova and Svetlik (1984) and Popovic (1987). However, foliar application of fertilizer at full development of leaf area generally increased seed yields in the former Czechoslovakia. Foliar applications increased number of seed heads, number of heads and 1000 seed weight (Voricek and Pouleh, 1986).

5.3. *Combine operations*

Seed yields are much lower in eastern U.S. than in the western U.S. Several factors are involved including lower numbers of flower heads per unit area, inadequate pollination, and smaller seed size. One factor sometimes overlooked is inefficient seed threshing equipment. A study was conducted in Kentucky (USA) in 1972–73 to determine the amount of seed lost in tailings during combine operations (Taylor, unpublished) . Tailings were saved and rethreshed to ascertain the amount of seed lost on the ground after passing through the combine (Table 15.2). These data clearly indicated that either with or without desiccation (paraquat), seed losses through tailings due to

Table 15.1
Results of applying Alar (Daminozide) on seed yield of red clover.

Location	Time of Application	Effect on Clover	Seed Yield Increase %	Reference
Ireland	25 to 40 days after cutting first crop	Increased head number	Variable, depending on treatment	Puri and Laidlaw, 1983
Russia	Early bud stage	Decreased lodging	23 to 29	Perepravo and Antonov 1984
Germany	Flower bud formation	Increased head number Reduced seed germination	17 to 37	Jakesova and Svetlik, 1984
Canada	First and second crop	Increased head number Increased seed setting	>100 for 2 of 4 cultivars, 0 for 2 cultivars	Christie and Choo, 1990
Poland	At start of flowering	Reduced plant height and 1000 seed wt. Increased head number	18 to 60	Rybak and Walczak, 1988
Russia	Flower bud information	Increased head number	17 to 38	Trofimova et al., 1987
France	Late bud stage	Decreased corolla length	5 (2x) to 69 (4x)	C. Mousett-Declas (personal communication)

improper combine adjustment were very high. With the desiccant, loss averaged 33% and without desiccant 40%. This does not include seed lost by shattering during the swathing process.

Table 15.2

Seed yields and losses through combine tailings with and without desiccant.

Year	Farmer	Yield kg/ha Desiccant[a]	% Yield loss	Yield kg/ha No desiccant	%Yield loss
1972	Grower 1	325	23	309	40
1973	Grower 2	143	57	143	55
1973	Grower 3	52	31	56	42
1973	Grower 4	46	21	39	25
1974	Grower 5	–	–	66	30
1973	Grower 6	–	–	290	59

[a] Paraquat applied; foliage cut and swathed before combining.

Fig. 15. 2. Red clover heads L, multiple parted head, R, normal head.

The effect of different combining treatments were examined in a study in 1972 (Table 15.3). Under these fairly weedy conditions, use of a desiccant and windrowing was beneficial but tailings losses were still significant. Data from other fields where weeds were not a problem indicated no advantage of desiccation over direct combining without desiccation. Considerable seed was lost by shattering during the windrowing process (not shown in Table 15.3). To prevent this loss and loss of seed through tailings, careful combine adjusting, harvesting seed in the rough (i.e., partial cleaning of chaff during the combine run) and direct combining is recommended.

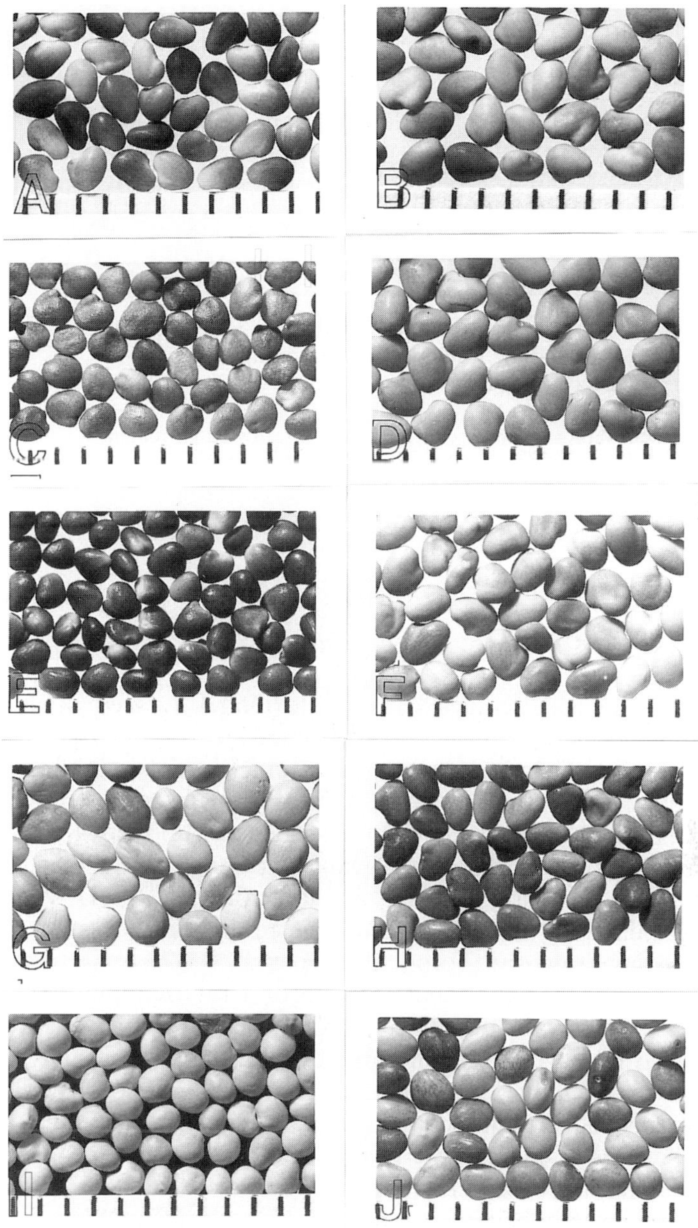

Fig. 15. 3. Seeds of red clover, and its close relatives. Marks are 1mm apart.
A - *T. pratense (diploid)* B - *T. pratense (tetraploid)* C - *T. diffusum* D - *T. medium* E - *T. pallidum* F - *T. ambiguum* G - *T. heldreichianum* H - *T. hirtum* I - *T. alpestre* J - *T. rubens.*

Table 15.3

Seed yields and losses through combine (tailings) in four treatments.

Treatment	Yield (kg/ha)	% Yield loss
Windrow	231	23
Desiccated (Direct)	326	23
Swath	267	33
Direct	309	40

Windrow - Mowed and windrowed with a haybine, and turned once (by hand) to facilitate drying.
Desiccated - Sprayed with paraquat and combined direct.
Swath - Mowed with haybine and allowed to remain until combined without pickup attachment.
Direct - No treatment - combined direct.

5.4. *Breeding*

One means of increasing seed yields is via breeding. Low seed yields mitigate against acceptance of red clover cultivars by seed growers even if forage yields are high. Therefore, it is important to determine the seed-yielding capacity of genetic stocks and to select those that are not inferior. Fortunately, in some cultivars, forage and seed yields are related. Smolikova et al. (1987) reported that the correlation between forage and seed characters in tetraploid cultivars and diploid genotypes was medium strong. In Russia, selection for high seed yields along with disease resistance and hardiness has been the focus of breeding programs (Voronchikhina et al., 1987).

The effect of phenotypic recurrent selection for multiple-parted flower heads on seed yields was examined in Kentucky (USA) (Fig. 15.2). After six generations of selection, the number of multiple-parted heads was 63% compared to 7% for the basic population. The number of seeds per head was greatly increased, but the number of heads was reduced because of inbreeding, and stem fasciation of the multiple-parted flower head genotypes. Seed yields per plant were not increased (Taylor et al., 1985).

Red clover normally has one seed per pod but occasionally, two seeds are developed. In Russia, the effect of selecting two seeded types on red clover seed yields was examined. The number of two seeded pods was correlated with number of seeds per inflorescence. After two cycles of selection, the selected population exceeded the standard in seed yield by 11.8 to 38.7% (Novoselova et al., 1988).

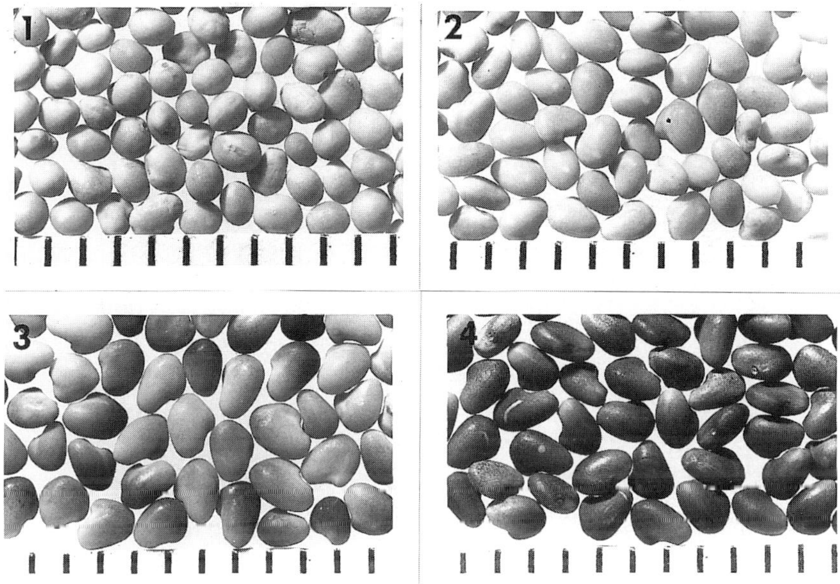

Fig. 15. 4. Variation in coloration of red clover seeds. Marks are 1mm apart.
1 - white 2 - yellow 3 - intermediate 4 - purple.

Table 15.4

Characteristics of seed of *Trifolium* species.

Species	Seeds per gram (no.)	Standard error	Color
T. pratense (2×)	626	19.3	yellow to purple
T. pratense (4×)	422	6.1	yellow to purple
T. pratense (WG)[a]	515	16.9	yellow to purple
T. diffusum	798	12.9	brown
T. pallidum	740	55.3	brown to purple
T. ambiguum	669	18.1	yellow brown
T. heldreichianum	578	45.0	yellow brown
T. alpestre	535	14.0	greenish brown
T. rubens	566	51.6	greenish yellow
T. medium	366	32.4	yellow
T. hirtum	318	20.0	yellow to reddish brown

[a] Seed produced in Washington State (USA); all others in Kentucky (USA), except *T. alpestre* produced in Madison, WI (USA).

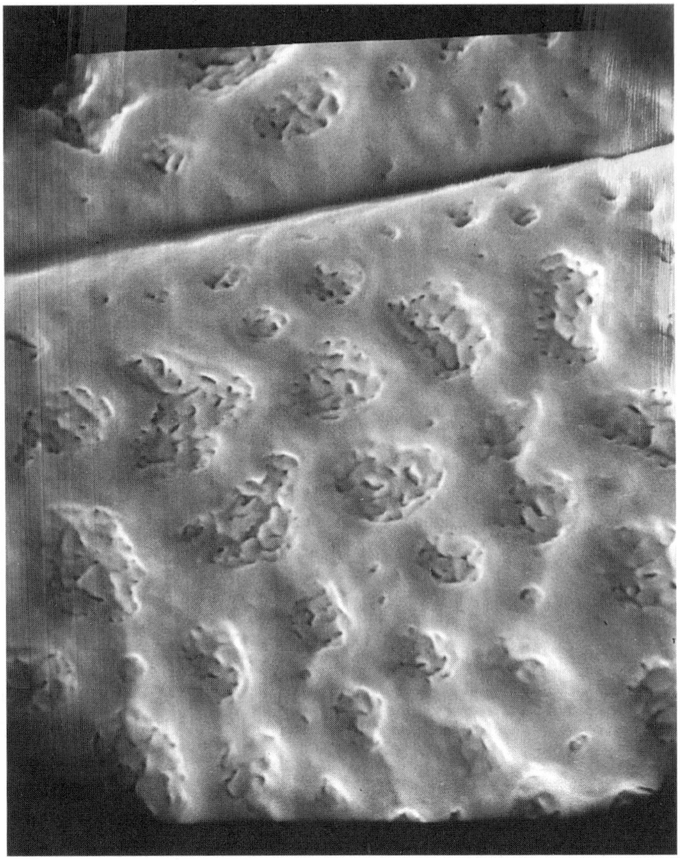

Fig. 15.5. Surface of a red clover seed as shown by the scanning electron microscope (6000×).

6. Seed characteristics

Seeds of red clover and its close relatives are approximately the same size (Table 15.4, Fig. 15.3). See Rincker and Rampton (1985) and Taylor (1987) for seed weights of other clover species. Red clover seeds should be sown at a depth of about 1 cm but greater depths may be successful in light or organic soils. Seeds of the same size may be difficult to separate with standard seed cleaning equipment. Difficulties in identification may also occur, particularly when seed of mixed species are harvested (during plant exploration, for example). To further confound the problem, seeds of red clover and other species may have different colorations and sizes due to circumstances of production. For example, seeds of red clover produced in Washington State (USA) are approximately 18% larger and more intensely colored than those produced in eastern U.S. Also, seeds of different ecotypes and cultivars of red clover may vary in color and size due to natural or artificial selection. The seeds of the cultivar Tennessee Purple Seeded are almost totally purple, whereas the seeds of the cultivar Kenstar are almost pure yellow and other seed colors occur in experimental strains (Fig. 15.4) (Quesenberry et al., 1991).

Fortunately, shape and some other characteristics remain relatively constant within species, and may be used as means of identification. Shape classifications (in one dimension) included ovate, obovate, elliptical, plump, and compressed. Other descriptive characteristics include the furrow, the hilum and notch, lateral margin, radical margin, micropyle, reniform (kidney shapes) and the strophiole (See Isley, 1948 for drawings illustrating these features). The scanning electron microscope may reveal differences in the seed coat otherwise unobservable (Fig. 15.5).

7. Seed storage

7.1. *Short-term*

Large amounts of seed that are transported in commerce prior to sale may be stored for up to three years at ambient temperatures. Although low temperature storage is desirable, it is often not possible due to seed volume. It is important that all characteristics of seed, particularly vigor, be optimized for storage. In New Zealand it was found that after 11 months, low-vigor seed lots even with an original germination percentage of 90% declined in germination to about 70%, whereas high vigor lots had not changed. Seed moisture content even for high vigor lots should be less than 10%. It is apparent that decisions about the length of time a seed lot can be stored safely should be based on vigor tests as well as germination percentage (Wang and Hampton, 1991).

7.2. *Long-term*

Small lots of seed such as breeder or germplasm lots are often maintained indefinitely. Seeds stored in a high nitrogen content atmosphere (98–99%) may be maintained successfully up to 6 years although germination percentage begins to decline after 3 years (Syroedov et al., 1987). However, red clover seed with high original germination (over 92%) maintained 90% germination for 13 to 20 years when stored at -5 to -15 °C. Low seed moisture and cold temperatures were more important than the type of atmosphere (Rincker and Rampton, 1985). This type of storage appears to be a safe procedure for red clover seeds. It also is probably safe for most if not all of the small seeded *Trifolium* species.

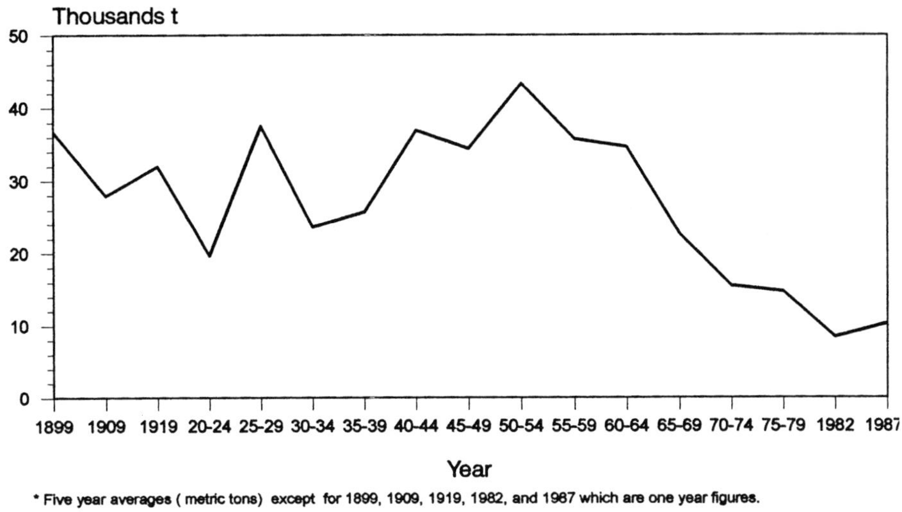

Fig. 15. 6. Red clover seed production in United States from 1899 to 1987.

8. Maintenance of genetic stability during seed multiplication

It is the responsibility of the originating plant breeder to provide adequate supplies of seed of a released cultivar, and to ensure that the characters bred into the cultivar are available when the seed is sold to farmers. For red clover, as well as for several other forages, seed yields are lower in the originating area than in areas of specialized seed production. In the U.S., the seed producing area is the western states, usually the Pacific Northwest. Genetic shifts to less persistent types have been shown to occur in red clover during seed increase from breeder to certified seed (Taylor et al., 1966, 1979, and 1990). The prevailing evidence suggests that earlier blooming types that produce more seed in the west than late blooming types are less persistent when regrown in the forage producing area. To correct this problem for the Kenland and Kenstar cultivars, seed production for certification was allowed only in northern locations in the Pacific Northwest, and the registered seed class has been eliminated thus reducing the allowable number of seed increases to two; foundation and certified. Recertification of seed from certified seed fields is prohibited. Strict adherence to the requirements of the Association of Official Seed Certifying Agencies concerning land history, volunteer seedlings, isolation, and seed cleaning also is recommended. Regulations for maintenance of genetic stability also exist in other countries, and may be even more strict than those in the U.S.

Table 15.5

Seed production (t) by countries at approximate dates of 1950, 1970, 1990.

Country	1950	Index [a]	1970	Index[a]	1990	Index[a]
USA	41,808	100	27,588	66	12,507	30
Canada	3779	100	5621	149	3901	103
New Zealand	1556	100	992	64	300	19
France	6900	100	5260	76	1165	17
Norway	301	100	15	5	107[b]	36
Sweden	5700	100	899	16	613	11
United Kingdom	—	—	1181	100	20	2

[a] Based on 1950 = 100 except United Kingdom where 1970 = 100.
[b] Average of 1987 to 1994 yields.

8.1. *Trends in red clover seed production*

The leading western states of the U.S., according to the 1987 U.S. Department of Agriculture Census (the latest available) in order of total amount of seed produced were Oregon, Washington, Idaho, and California. These four western states produced over 5984 metric tons (t) which was about 58% of the total USA seed production. Seed yields ranged from 356 (Oregon) to 918 (California) and averaged almost 435 kg per hectare.

Total red clover seed production has declined recently (Fig. 15.6, Table 15.5). For many years, total production of clean seed ranged from about 35,000 to 45,000 t, reaching a maximum in 1950. However, production which in 1964 was about 34,617 t, has declined steadily to about 8455 t in 1982 and 10,288 t in 1987.

In 1974, eight states, Missouri, Illinois, Minnesota, Ohio, Michigan, Wisconsin, Indiana, and Iowa (termed eastern hereafter) produced about 7746 t of seed, or about 60% of the U.S. production. The percentage declined to 44 in 1978, and to 38 in 1987, indicating a shift to western U.S. Although the total production of these eight states has declined, it represents an increasingly larger proportion of the eastern production ranging from 78 percent in 1974 to 91% in 1987. On the other hand, seed yields in kg per hectare have increased in the eastern states from 97 in 1974 to 124 in 1987. Eastern seed production has likewise decreased from about 100,000 ha in 1974 to about 34,000 ha in 1987. Corresponding western figures are about 8,000 ha in 1974 and almost 13,500 ha in 1987. This shift probably means that fewer hectares of red clover are being sown for forage production in eastern and central U.S.

A survey of personnel in several red clover seed producing countries indicated that seed production also has declined since reaching a maximum about

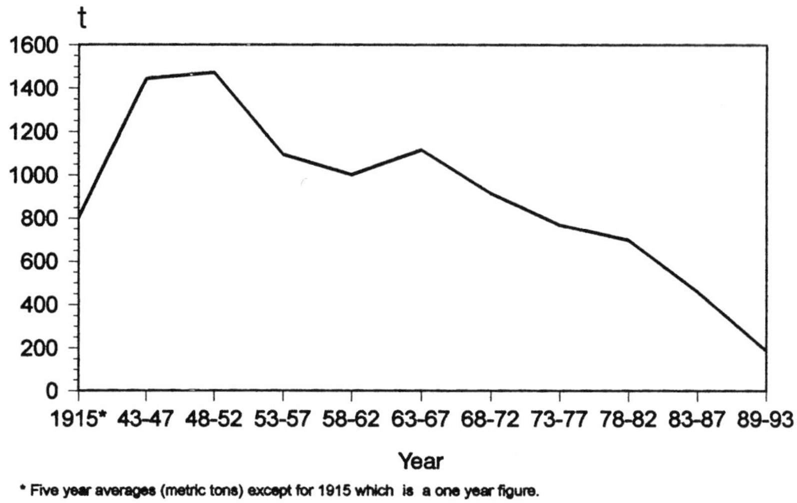

Fig. 15. 7. Red clover seed production in New Zealand from 1915 to 1993.

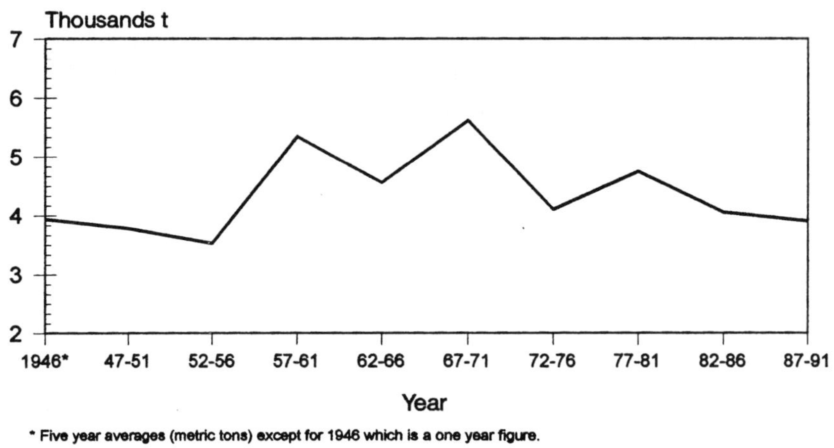

Fig. 15. 8. Red clover seed production in Canada from 1946 to 1991.

1950 (Table 15.5). The causes of this decline are varied but generally are related to declining use of red clover as a forage. In New Zealand, production of seed for exports was important in the decade from 1945 to 1955, and decreased somewhat but was still important from 1955 to 1975 (Fig. 15.7). Since then, declining exports and declining usage of red clover, and increased usage of white clover decreased the demand for red clover seed (A.V. Steward, W. Rumball, personal communication).

Likewise, Sweden is producing only a fraction of the amount produced in 1950. This decline is a result of several factors: (1) amount of red clover seed sown in mixtures from up to 50 to 70% to as low as 25 to 30%; (2) reduction in total forage area; (3) use of grasses alone has increased; and (4) reduction in exports (B. Anderson, personal communication). In France, seed production is quite variable depending upon exportation and variation in stocks. Nevertheless, 5-year production averages clearly show the decline (Table 15.5). French production in 1990 was only 22% of the production in 1970 (C. Mouset-Declas, personal communication). Other countries showing declines include Norway and the United Kingdom (Table 15.5). Countries in which the production, never large, is now virtually zero include Belgium, the Netherlands, Japan, and Brazil. No information was obtained from Russia, Poland, Czechoslovakia and Romania and it is possible that declines, if any, are not as great in these countries.

Only in Canada is there information that red clover seed production is not decreasing significantly (Fig. 15.8). Canada produces both single cut (mammoth) and double cut types. Starting about 1955, seed production of the single cut type has exceeded the double cut type and today comprises about 75% of the production. Doubtless, a large part of this seed is exported (T.M. Choo and R. Michaud, personal communication).

Thus it is apparent that in most countries red clover seed production has declined as use of red clover for forage has decreased. See Chapter 17, Future Trends for further discussion of the factors involved.

References

Bawolski, S. and H. Scibior. 1985. Effect of sowing rate on the seed yield and yield components of a diploid red clover. *Pamietnik Pulawski.* 85:117–127.

Belzile, L. 1990. Influence of cultivar and vegetative stage of cutting on seed production of red clover. *Canad. J. Plant Sci.* 70:1071–1080.

Belzile, L. 1994. Potential yield of red clover seed in the seedling year. *Canad. J. Plant Sci.* 74:807–809.

Benkov, B. and Ts. Dimitrova. 1983. Chemical control of weeds in fields of red clover for seed and forage. *Rasteniev dni Nauke.* 20:24–30.

Binderva, A. 1985. Seed production of red clover in relation to the date of the first cut. *Sbornik Vedeckych Praci, Oseva, Vyzkumny a Slechtitelsky Ustav Picninarsky, Troubsko u Brna.* 9:129–136.

Bojas, Z. 1986. Effect of new herbicides on weed eradication in red clover stands. *Pol'no-haspodarstvo.* 32:784–791.

Bydrina, O.S. and V.G. Voskresenskaya. 1980. Chemical control of weeds in clover for seed production. *Zashchita kormovykh kultur ot vreditelei, boleznei i sornyakov.* 1980:107-109.

Christie, B.R. and T.M. Choo. 1990. Effects of harvest time and Alar-85 on seed yield of red clover. *Canad. J. Plant Sci.* 70:869–871.

Fairey, D.T., L.P. Lefkovitch and J.A.C. Lieverse. 1989. The leafcutting bee, *Megachile rotundata* (F.): a potential pollinator for red clover. *Zeitschrift fur Angewandte Entomologie.* 107:52–57.

Isley, D. 1948. Seed characteristics of the common clovers (Trifolium). *Iowa State Coll. J. Sci.* 23:125–136.

Jakesova, H. and V. Svetlik. 1984. Influence of sowing, trace elements, and growth regulators on yield and quality of seed of tetraploid red clover. *Wissenschaftliche Beiträge,* Martin Luther Universitat Halle Wittenberg. 54:378–389.

Macfarlane, R.P., H.J. van den Eude and R.P. Griffin. 1991. Pollination needs of 'Grasslands Pawera' red clover. In C. van Heemert and A. Ruiyter (eds.), *The Sixth International Symposium on Pollination.* pp. 399–404. Tilburg, The Netherlands.

Macfarlane, R.P. and R.P. Griffin. 1985. An assessment of bumble bees and honey bees as pollinators of red clover. In R.B. Chapman (ed.), *Fourth Australasian Conf. on Grassl. Invert. Ecology.* pp. 114–120. Lincoln College, Canterbury, N.Z.

Novoselova, A.S., E.A. Kositsyna-Pinegina and N.V. Razgulyaeva. 1988. Form of red clover with two-seeded pods. *Selektsiya i semenovodstvo. Moscow.* 1988:27–28.

Oliva, R.N., J.J. Steiner and W.C. Young, III. 1994a. Red clover seed production: I. Crop water requirements and irrigation timing. *Crop Sci.* 34:178–184.

Oliva, R.N., J.J. Steiner and W.C. Young, III. 1994b. Red clover seed production: II. Plant water status on yield and yield components. *Crop Sci.* 34:184–192.

Perepravo, N.I. and V.I. Antonov. 1984. Effect of growth regulators on seed yield of red clover. *Khimiya v Sel'skom Khozyaistve.* 11:48–49.

Popovic, I. 1987. Stimulation of nectar production of red clover by means of microelements. *Vedecke Prace Vyzkumneho Ustavu Vcelarskeho v Dole.* 9:167–177.

Puri, K.P. and A.S. Laidlaw. 1983. The effect of cutting in spring and application of Alar on red clover (*Trifolium pratense* L.) seed production. *J. Appl. Seed Prod.* 1:12–18.

Puri, K.P. and A.S. Laidlaw. 1984. The effect of time of harvest on seed production of three red clover cultivars. *Grass and Forage Sci.* 39:221–228.

Quesenberry, K.H, R.R. Smith, N.L. Taylor, D.D. Baltensperger and W.A. Parrott. 1991. Genetic nomenclature in clovers and special-purpose legumes: I. Red and white clover. *Crop Sci.* 31:861–867.

Rincker, C.M. and H.H. Rampton. 1985. Seed production. In N.L. Taylor (ed.), *Clover Science and Technology.* ASA-CSSA-SSSA Mono. 25:417–443. Madison WI.

Ruszkowski, A., M. Bilinski, A. Kosior, J. Bok and K. Kaczmarska. 1989. The bumble bees of the Malopolska uplands (Poland). *Pszczelnicze Zeszyty Naukowe.* 33:33–54.

Rybak, H. and S. Walczak. 1988. Effect of growth regulators on red clover seed yields. *Roczniki Nauk Rolniczych, E Ochrona Roslin.* 17:227–235.

Skuterud, R. 1982. Weed control in the year of sowing. Engfroavl Roskilde, Denmark; Statens Forsogsstation 1982:190–196.

Smolikova, M., J. Kailerova and B. Nedbolkova. 1987. Correlation between fodder and seed traits in genotypes of red clover. *Sbornik Vedeckych Praci Vyzkumneho a Slechtitelskeho Ustavu Picninarskeho v Troubsku u Brna.* 10:83–92.

Syroedov, V.I., V.D. Nadykta, N.I. Nazasova, M.S. Ragulin and G.S. Gorshkov. 1987. Storing the seeds of perennial herbage plants in a controlled atmosphere with an increased nitrogen content. *Doklady Vsesoyuzoi Ordena Lenina i Ordena Trudovogo Krasnogo Znameni Akademii Sel'skokhozyaistvennykh Nauk Imeni V.I. Lenina.* 1987:18–20.

Taylor, N.L., M.K. Anderson and D.M. Tekrony. 1972. Producing red clover seed in Kentucky. *Univ. of Kentucky Coop. Ext. Service Leafl.* AGR 2:1–4.

Taylor, N.L., P.L. Cornelius and M.G. Long. 1985. Phenotypic recurrent selection for multiple-parted flower heads in red clover. *Crop Sci.* 25:489–494.

Taylor, N.L., E. Dade and C.S. Garrison. 1966. Factors involved in seed production of red clover clones and their polycross progenies at two diverse locations. *Crop Sci.* 6:535–538.

Taylor, N.L., R.O. May, A.M. Decker, C.M. Rincker and C.S. Garrison. 1979. Genetic stability of Kenland red clover during seed multiplication. *Crop Sci.* 19:429–434.

Taylor, N.L. 1987. Forage legumes. In W.R. Fehr (ed.), *Principles of Cultivar Development, Crop Species.* Vol 2:209–248. Macmillan Publ. Co. New York.

Taylor, N.L., C.M. Rincker, C.S. Garrison, R.R. Smith and P.L. Cornelius. 1990. Effect of seed multiplication regimes on genetic stability of Kenstar red clover. *J. Appl. Seed Prod.* 8:21–27.

Trofimova, T.A., N.I. Perepravo and O.P. Banadyseva. 1987. The use of retardants for increasing cold resistance and seed yield of red clover. *Sbornik Nauchnykh Trudov, Vsesoyuznyi Nauchno issledovatel' skii Institut Kormov.* 35:142–145.

Voricek, Z. and Z. Pouleh. 1986. Foliar fertilizer application to red clover seed crops. *Rostlinna Vyroba.* 32:863–872.

Voronchikhina, A.P., L. M. Chetvertnykh and M.I. Koromyslova. 1987. Early red clover hybrids for breeding for seed production. *Selektsiya, semenovodstvo i tekhnologiya vozdelyvaniya mnogoletnikh trav v Severo Zapadnoi oblasti Necherozemnoi zony* RSFSR. 1987:40–43.

Wang, Y.R. and J.G. Hampton. 1991. Seed vigor and storage in Grasslands Pawera red clover. *Plant Varieties and Seeds.* 4:61–66.

Wilczek, M. 1986. Effect of sowing dates and rates on yields of red clover (*Trifolium pratense* L.) seed. *Biuletyn Instytutu Hodowli i Aklimatyzacji Roslin.* 159:57–63.

16. Germplasm Acquisition, Maintenance, and Evaluation

1. Introduction

Genetic resources are the foundation of any crop improvement program. Red clover has spread around the globe through contact of people from the Middle East and Eastern Europe with other countries in the Old World and the New World. It is thus difficult to identify the "area of origin" of red clover. Red clover is thought to have originated in Asia Minor and southeastern Europe, but it was already cultivated in Europe in the third and fourth century (Taylor 1973, 1985). Red clover was recorded in Italy by 1550, France by 1585, England by 1645, and the USA by 1663. The early spread of clover was likely primarily by seed on mature heads in hay and bedding. Piper (1924) believed that red clover had a greater influence on civilization than any other forage plant. As germplasm of red clover was introduced by settlers to new lands, natural selection has resulted in the development of numerous well adapted and highly productive ecotypes and farmer landraces.

This selection for local ecotypes of red clover has, however, presented problems for taxonomic description of the species. The traditional taxonomist is bewildered by the immense variation found within species of cultivated plants (Harlan, 1975). This truth is borne out by the extreme polymorphism of red clover for which no less than 40 binomials have been published (Zohary and Heller, 1984) . These authors further state that there is little experimental evidence to testify to the constancy of traits on which various species, subspecies, and varieties have been erected. Nevertheless, they recognize six botanical varieties based primarily on hairiness, stem length, multiple parted heads, and stem branching. Although some of these taxa may be merited, recent breeding research has shown that plant hairiness (Baker et al., 1992), stem length (Bowley et al., 1984), and multiple parted heads (Taylor et al., 1985) are intraspecific traits easily manipulated by recurrent selection breeding.

It becomes obvious that enormous variability exists in red clover germplasm. Nevertheless, in recent years development and planting of improved cultivars has resulted in a decline in the number of farmer ecotypes and wild forms in known centers of diversity. This erosion of genetic diversity is a concern even in the clovers (Taylor et al., 1977). The primary concern thus becomes how to collect maintain, evaluate, and utilize this currently existing germplasm for the improvement of red clover production. In Chapter 12, we have proposed that any breeding program should first concentrate on selection from recently developed cultivars and local populations. If the desired traits are not found there, and cannot be increased by recurrent selection, then plant introductions should be evaluated.

Plant introductions may be of more value in the earlier stages of breeding and selecting cultivars of red clover adapted to a new region. An example of this is reflected in the pedigree of Cherokee, a cultivar developed for the lower southeastern United States, an area where red clover previously had not been grown. The base population on which selection for Cherokee was initiated consisted of a combination of about 75% selections from plant introductions and 25% from two older cultivars (Quesenberry et al., 1993). The selected plant introductions in this population previously had been identified as having early vigor and reseeding potential over a three year period. The base population was then subjected to five cycles of selection (one cycle per year) for early vigor when grown as spaced plants in a root-knot nematode infested field. This selection for early vigor apparently resulted in concurrent selection for a reduced dormancy response to shortening days. Individual plants in Cherokee will flower normally on days of less than 12 hours, whereas cultivars developed in the central and northern U.S. and in Europe generally exhibit a strong dormancy response to shortening days in fall, and will usually remain vegetative on day lengths of less than 12 hours. This reduced dormancy results in greater early season growth in mild climates compared to cultivars selected at more northern latitudes. Additional specific selection in Cherokee for root-knot nematode resistance has resulted in development of germplasm with high levels of root-knot nematode resistance (Call and Quesenberry, 1994).

Reports from Russia also illustrate the value of red clover plant introductions and older cultivars in development of improved cultivars. The impact of N. I. Vavilov's efforts to collect diverse germplasm, including red clover introductions, on the development of Soviet red clover varieties has been documented (Kuleshov et al., 1987) . Of 54 varieties reviewed, 68.5% were produced using collections of local cultivated populations, 9.2% were produced using wild populations, and 16.6% were produced from previously developed Soviet varieties, while only 5.5% were produced from foreign sources. Likewise, in the Ukraine a study of red clover introductions from the Vavilov Institute of Plant Industry (VIR) identified 26 lines and hybrids which exceeded released cultivars in yield of seed and forage (Drach et al., 1988). The diploid cultivar Darunok and the tetraploid Kumach were developed from this material. The earlier flowering cultivar Kumach had strong winter persistence and a vigorous root system. Use of introductions from VIR in selecting for winter persistence, resistance to *Sclerotina trifoliorum*, and dry matter yield in Vologda province has also been reported (Ryabova, 1986) . Previously developed cultivars were the source of resistance to *Sclerotinia trifoliorum, Kabatiella caulivora, Pseudopeziza trifolii,* and *Erysiphe communis* f.sp. *trifolii (Microsphaera trifolii* var *trifolii)* in recently developed cultivars. Selections for resistance to these pathogens were carried out under natural infestation in Leningrad (St. Petersburg) province (Drozdova, 1988).

Another example of the use of locally adapted material is that of Banciu and Tucra (1982) from Romania. They reported that local populations generally had high yields and served as good sources for additional breeding and selection. The red clover cultivars Napoca-Tetra and F8 were developed from this program.

2. Germplasm collections of red clover

This section attempts to document and summarize the major red clover germplasm collections known to exist. Certainly, collections held by individuals and even some country collections may have been overlooked. Table 16.1 is a listing of collections held at various locations. Where the information was available, the holdings have been divided into the categories of 1) advanced breeder lines, 2) primitive cultivars and landraces, and 3) ecotypes (semi-natural or wild). If the publications did not classify type of accessions, they were listed as unclassified. Data included in this table came from " A Guide to European, Forage Databases" (IBPGR, 1991), "Directory of Germplasm Collections 7. Forages . . ." (Bettencourt et al., 1992), access to the U.S. National Plant Germplasm System (NPGS), and personal contact with curators of various germplasm collections. The International Plant Genetic Resources Institute (IPGRI), formerly IBPGR, has coordinated the European Cooperative Program for the Conservation and Exchange of Crop Genetic Resources (ECP/GR) since 1981. The Forage Working Group first met in 1984. The publication of "A Guide to the European Forage Databases" (IBPGR, 1991) lists the red clover accessions held by various European countries. The red clover coordinator for ECP/GR is located at Station Fe'de'rales de Recherches Agronomiques de Changins, Route de Duiller, CH-1260 Nyon, Switzerland. A computerized listing of the holding in the various European collections is available from that office.

Among the larger collections is the U.S. NPGS collection of perennial *Trifolium* spp. including red clover that has been maintained at the Northeast Regional Plant Introduction Station at Geneva, New York. Recently, this collection was moved to the Western Regional Plant Introduction Station, 59 Johnson Hall, Washington State University, Pullman, WA 99164–6421. Annual *Trifolium* species in the U.S. collection are maintained at the Southern

Table 16.1

Red clover germplasm accessions classified by type and country where accessions are maintained.*

Maintaining Country	Type of accession: Adv. Cultivars & Brdr. lines	Prim. cultivars & landraces	Ecotypes (wild)	Unclassified	Total
Belgium	3	7	2		12
Bulgaria	5		1		6
Switzerland[1]	7	8			15
Canada			118		118
China(Mongolia)				72	72
Czechoslovakia[1]	296	33	48	203	580
Germany[1]	67	37	32		136
France[1]	82	103		481	666
Greece[1]	106		25		131
Hungary	37	211	132		380
Italy			9		9
Japan				437	437
Netherlands	15		121		136
New Zealand	490		679		1169
Poland				60	60
Romania[1]			10	224	234
Russia				2913	2913
South Africa[1]	10	5		29	44
Sweden	47	122	131		300
Syria				55	55
Turkey			60		60
United Kingdom[1]	24	7	26	48	105
United States[2]	595		640		1235
Total	1784	533	1916	4640	8873

*Data synthesized from "A Guide to the European Forage Databases, (IBPGR, 1991) and Directory of Germplasm Collections (IBPGR,1992). Data from other countries obtained by personal contact with: Dr. W. M. Williams, Curator, Margot Forde Forage Germplasm Center, Palmerston North, NZ. U.S. data from Germplasm Resources Information Network.

[1] Combined data from two or more institutes within a country.

[2] Data not easily separated between advanced cultivars and primitive landraces.

Regional Plant Introduction Station at Griffin, GA. All *Trifolium* accessions in the U.S. system can be accessed through the Germplasm Resources Information Network (GRIN). Access to GRIN can be obtained by contacting the Germplasm Resources Laboratory, Building 001, Room 127, BARC-West,

Beltsville, MD 20705. Red clover accessions for New Zealand and Australia are maintained at the Margot Forde Forage Germplasm Centre, Private Bag 11008, Palmerston North, New Zealand. Another major collection of red clover is that held at the Vavilov Institute for Research (VIR) in St. Petersburg, Russia.

A review of the data assembled in Table 16.1 reveals that of the 4,233 accessions classified by type, about 45% of accessions are wild ecotypes, 13% are primitive landraces, and the remaining 42% are advanced cultivars. Assuming that 4640 unclassified accessions fall in these same proportions, we estimate that wild germplasm accessions number about 4000. Lack of classification data on the large collection at VIR causes the above number to be somewhat subjective. There is likely a high degree of duplication of cultivars between the European, U.S., and New Zealand collections.

Seed storage is a major concern in germplasm conservation. Red clover seed with good original germination over (92%) was shown to maintain greater than 90% germination for 13 to 20 years when stored at -5 to -15 °C (Rincker, 1983). Even frequent thawing of seed maintained at subfreezing temperatures had little effect on viability of red clover seed (Rincker, 1974). Thus, the recommended conditions for red clover seed storage are 10% or lower moisture stored at -18 °C. The U.S. National Seed Storage Laboratory at Ft. Collins, CO maintains long term samples of U.S. red clover introductions under liquid N at -196 °C. No long term studies comparing storage of red clover seed at -5 to -15 vs. liquid N temperature were found, but one experiment in progress shows no difference in germination of clover seed stored at -18 versus -196 °C for 10 years (P. Stanwood, personal communication). In related work with *Allium*, seed vigor tests have shown more reduction in seed vigor when stored at -18 than at -196 °C.

3. Acquisition of red clover germplasm

The three primary means of acquiring any type of germplasm are: 1) personal contact exchanges, 2) exchanges among gene banks, and 3) collecting expeditions. Since the latter is most expensive, the first two are preferred. With the recent coming into force of the Convention on Biological Diversity, all types of germplasm exchange may come under more scrutiny. The issue of the status of *ex situ* collections established before the Convention became effective remains to be settled, but hopefully few or no limitations will be placed on use of these resources.

National gene banks often exchange parts of collections with other national gene banks to insure backup of important collections. However, the cost of maintenance of accessions and, particularly the cost of seed increase for red clover accessions, must be carefully considered in such exchanges. Care should also be taken to avoid duplication of previously held accessions. For example, a review of red clover holdings in the U.S. NPGS shows that more than 75% of the red clover cultivars in the collection have passport data of foreign origin, and are likely duplicated in European, Russian, or New Zealand collections. Indiscriminate exchange of germplasm between collections is of little value and may result in poor maintenance of more valuable accessions due to financial constraints from attempting to maintain large numbers of accessions.

With the large number of accessions currently held in various red clover gene banks, any effort to collect new germplasm should be examined thoroughly to avoid duplications. An effective strategy for germplasm collection is to identify desirable traits, e.g., low pH tolerance and areas where these traits potentially exist. Ecogeographic classification and maps may be helpful in this process.

A major project in the IPGRI Genetic Diversity Program is focused on identifying collecting strategies which will increase the genetic diversity located in, and collected for, *ex situ* conservation (IPGRI, 1993). Genetic variation is well known to be associated with eco geographic diversity (Porceddu and Dowania, 1992). Thus, conscious efforts should be made to include the full range of ecologically relevant variation in efforts to preserve germplasm in *ex situ* collections (Huenneke, 1991). Furthermore careful consideration of sampling strategies based on species and population distribution across the targeted landscape is critical when developing a collecting scheme (Brown and Briggs, 1991). Recent availability of computer based Geographic Information System (GIS) technology may enhance effectiveness of collection planning (Steiner and Poklemba, 1994).

Procedures for preparation prior to undertaking a clover germplasm collecting expedition and details for collecting and documenting collections have been outlined by Gillett and Smith (1985). Also recommended for study prior to undertaking a germplasm collecting expedition is "Crop Genetic Resources Field Collection Manual" (Hawkes, 1980). This manual provides numerous insights into steps for planning and execution of a germplasm collection expedition. One new tool that has recently become available to assist with geographic location documentation is the Global Position Device (GPD). This instrument allows for extremely accurate determination of latitude and longitude; however, a barometric altimeter may be needed to supplement altitude readings of some GPDs. Collection of herbarium specimens and sampling of rhizobium nodules are desirable goals for clover collection. Nodule samples may be difficult to obtain in dry soils and at advanced stages of maturity when seeds are ripe. Herbarium samples usually are difficult to carry across international borders. Voucher samples should always be prepared when collections are grown out to verify field identification.

A most critical factor to the success of any germplasm collection expedition is the cooperation of local individuals who can facilitate language barriers, local customs, currency exchanges, and other personal items. Knowledge

Table 16.2

Germplasm collections expeditions for *Trifolium* species since 1977.

Year	Country	Region(s)	Total *Trifolium* Accessions	Red clover Accessions
U.S. NPGS sponsored				
1977	Greece, Crete, Italy	Entire country	254	39
1984	Romania	Entire country	37	7
1988	Yugoslavia[a]	Croatia, Bosnia, Slovenija	95	17
1990	Bulgaria[b]	South-central and east	148	5
1993	Bulgaria[b]	South-west and west	110	4
1993	Rep. Georgia[c]	Central and southern Caucasus Mts.	113	9
1994	U.S.A	California	63	0
1994	U.S.A[d]	Oregon	43	0
Australia sponsored				
1984	Turkey	North-east	52	13
1994	Morocco, Tunisia[e]	Various areas	39	9
New Zealand sponsored				
1986	S. W. Europe[f]	Italy, Spain, Portugal	370	52
1989	Southern USSR[g]	Caucasus Mts. of Georgia, Armenia, and Azerbaijan	237	38

[a]Included IBPGR sponsored cooperator.
[b]Included Bulgarian IIPGR sponsored cooperators.
[c]Included Republic of Georgia cooperators.
[d]Included New Zealand sponsored cooperator.
[e]Included U.S. sponsored cooperator.
[f]Joint sponsorship by France and IBPGR.
[g]Joint sponsorship by U.K., USSR, IBPGR.

by these persons of specific locations where selected species can be found is perhaps more important than any other factor. At least twelve collection expeditions focused on *Trifolium* species have been conducted since the early 1980's. Eight of these were partially sponsored by the U.S. NPGS (Taylor and Rumbaugh, 1986; Taylor and Smith, 1990; Quesenberry and Smith, 1990; Pederson and Quesenberry, 1994, Smith and Hoveland, 1994). Two expeditions sponsored by New Zealand and two by Australia are also known (Forde and Easton, 1986). Reports from all these efforts bear out the importance of local cooperators as a part of any collection effort. A summary of the results of these collection trips is included in Table 16.2 and a brief review of the more recent collection expeditions will be presented.

Fig. 16. 1. Collecting annual and perennial *Trifolium* species in Bulgaria 1990.

A germplasm collection trip to the warm-temperate areas of Portugal, Spain, and Italy was jointly sponsored by New Zealand, France, and IBPGR for a two month period in June and July 1986. Areas visited included Spain: Andalucia (S.W.), costal Galicia, Asturias, and Cantabrica (N.W.); Portugal: Algarve (S.), along the western coast and in central and northern regions, Galicia, Santiago, La Coruna (N.E.); Italy: Unbria and Tuscany (C.), around Lodi (Po Valley) and along N.E. mountain slopes. Collections were made mainly from cut, grazed, or wild meadows and included the genera *Lolium, Festuca, Holcus, Trifolium, Lotus, Ornithopus, Medicago,* and *Vicia* for a total of 1244 accessions (Forde and Easton, 1986).

An expedition to collect *Trifolium* species and other forage germplasm in Romania was sponsored by the U.S. NPGS during the period 5–23 August 1984 (Taylor and Rumbaugh, 1986). The trip was conducted in cooperation with the Academy of Agricultural and Forest Sciences, Buchurest, Romania. Only thirteen different *Trifolium* species were collected including three annuals, although various flora list up to 40 species found in Romania.

A *Trifolium* species germplasm collection trip to the former Yugoslavia was sponsored by the U.S. NPGS in late August 1988 (Taylor and Smith, 1990). This effort also included a cooperator sponsored by IBPGR. The expedition was able to visit only three of the six Republics (Croatia, Bosnia, and Slovenija). The expedition collected 95 accessions over a period of 18 days. Seventeen accessions of red clover from diverse locations were obtained on this trip. Several of the areas where collections were made are currently inaccessible due to civil unrest, and it is possible that collection sites are severely disturbed. This illustrates the need for timely collections when contacts between governments are stable.

Fig. 16. 2. Collecting perennial *Trifolium* species in mountain meadow that is being hand mowed for hay (Romania).

A U.S. NPGS sponsored germplasm collection expedition to central and southeastern Bulgaria was carried out jointly with staff of the Institute of Introduction and Plant Genetic Resources (IIPGR), Sadovo, Bulgaria during the period of 14–27 July 1990 (Quesenberry and Smith, 1990). A preliminary trip by IIPGR in 1989 was extremely valuable in selecting dates for collection and potential collection sites. Due to seasonal variation almost all annual species were mature with dried heads in 1990 by the dates selected based on the 1989 survey, but this early season allowed collection of mature seeds from most perennial species encountered (Fig. 16.1). A total of 148 accessions, including five accessions of red clover from diverse habitats were collected. The flora of this area of Bulgaria has an abundance of *Trifolium* species that were relatively easy to collect in most areas visited. With most land still under centralized governmental control, grazing pressure was limited and seed were generally easy to collect.

A second joint U.S. NPGS and IIPGR expedition to the mountainous south central and western areas of Bulgaria was carried out during the period of 30 July to 13 August, 1993 (Pederson and Quesenberry, 1994). The timing was such that seeds of most perennials were mature, but several natural meadows were in the process of being harvested for hay (Fig. 16.2). With changes in land tenure, and increased livestock populations, overgrazing was a more significant limitation to collection in 1993 than 1990 (Fig. 16.3). A total of 110 collections representing 48 *Trifolium* species including four accessions of *T. pratense* from diverse habitats were obtained.

A U.S. NPGS collection expedition to the Republic of Georgia was conducted jointly with the Republic of Georgia Academy of Sciences from 27 July to 11 August 1993 (Smith and Hoveland, 1994). The U.S. participants

Fig. 16. 3. Overgrazing by sheep and goats of native pastures containing *Trifolium* species.

Fig. 16. 4. Plant explorers collecting *T. longipes* in Douglas Co. Oregon, Cascade mountains.

state in their report that the "expedition . . . would not have been possible without the cooperation of the Georgia Academy of Sciences", again stressing the value of local cooperators to collection success. This trip collected 113 samples of at least 12 different *Trifolium* species including nine accessions of red clover from diverse ecozones.

An expedition for collection of *Trifolium* species native to California, Oregon, and Washington states was sponsored by the U.S. NPGS and the Margot Forde Forage Germplasm Center, Palmerston North, New Zealand, during summer, 1994. This trip represents another good example of international cooperation on development and execution of germplasm collection expeditions. The collection effort was divided into two components with the first focused primarily on annual *Trifolium* species in California and included only the U.S. participants. Populations of annual and perennial (but especially perennial) *Trifolium* species in the western U.S. exist in very narrowly defined and often isolated habitats (Fig. 16.4). As stated by Gillett and Smith (1985), the distances involved are often great and along roads that permit only slow travel. The need for local "experts" was demonstrated in California and Oregon. Various U.S. Forest Service and U.S. Bureau of Land Management Botanists/Naturalists personnel as well as botanists from the Herbaria at the University of California, Davis, Oregon State University, Corvallis, and the Berry Botanical Garden, Portland, OR, directed the collecting team to many collection sites. The two trips resulted in 106 collections representing at least 38 different *Trifolium* species. Of these, 16 were not previously represented in the U.S. collection and 5 were listed by the U.S. Fish and Wildlife Service as possible candidates for threatened or endangered status. Permits to allow collection of threatened or endangered species were obtained from appropriate state and federal agencies. Attempts are being made to identify potential *in situ* preserves for some of these species, but a need for preservation in germplasm banks also exists. No accessions of red clover were collected, as primarily roadside populations were seen and these were assumed to be recently introduced as a result of the large volume of commercial red clover seed production in these states. All accessions were divided between the U.S. NPGS and the Margot Forde Germplasm Center, New Zealand.

4. Germplasm maintenance

Optimum conditions for long term storage were discussed above. The primary other consideration with maintenance of red clover germplasm is the need for seed stock regeneration. Since red clover is cross pollinated, each individual accession must be regenerated in isolation to insure maintenance of genetic integrity. Where many accessions are being stored (and ultimately may need to be regenerated) the cost of such increase can be large. It has been estimated that bee isolation cage increase of red clover accessions costs between U.S. $300 to $500 per accession. Other considerations for seed regeneration of red clover are: 1) a minimum of 100 individuals should be grown to represent the genetic diversity of an accession and to restrict inbreeding; 2) care should be taken to insure that all plants flower and contribute to the increase; 3) care should be exercised to prevent accidental seed contamination between accessions during the harvesting and cleaning process.

One tool which has been suggested to assist germplasm managers deal with maintenance costs of large numbers of accessions is the use of core subsets of accessions. This concept has been espoused by Frankel (1984) and Brown (1989a, 1989b) as a severely limited sample which represents the genetic diversity of a crop species and its relatives. Various authors have discussed the advantages and pitfalls of core subsets. The U.S. National Plant Germplasm System has developed general guidelines and procedures for developing core subsets. A symposium was held at the 1994 Crop Science Society of America annual meeting on the topic of "Using Core Subsets for Efficient Germplasm Evaluation and Management". Molecular markers may become an important tool for accessing the level of genetic diversity in germplasm collections (Kresovich et al., 1994).

The Clover and Special Purpose Legume Crop Germplasm Committee of the U.S. National Plant Germplasm System is charged with providing advice to U.S. germplasm managers on various issues including maintenance. This committee has developed and approved lists of accessions constituting core subsets of various *Trifolium* species. This committee has recently reviewed the U.S. collection of red clover accessions and recommended a core subset. The procedure used in identifying similarity groups of accessions of red clover was recently described (Kouame and Quesenberry, 1993). They primarily used historical morphological, agronomic, and disease response data on 15 descriptors to conduct a cluster analysis. This analysis identified 10 different similarity groups. The preliminary core contains at least one accession from every country of origin in the collection and at least two accessions from each cluster group. A few accessions with known attributes of pest resistance were added to the core. Core subsets should be fluid and should be adjusted when the range of variability for traits is extended or when new traits are evaluated (Brown, 1989a).

5. Germplasm evaluation

Germplasm accessions are of little value if they are only preserved in a germplasm repository. Systematic evaluations of accessions using recognized descriptors can facilitate identification of accessions with desirable attributes. List of descriptors for forage legumes including red clover have been published (IBPGR, 1984). Most germplasm collections use this list of descriptors or some modification of it. When evaluating accessions, it is important to sample a large enough number of individuals to identify variability which may be present at low frequency. Ideally, a minimum of 100 individuals should be evaluated of each accession; however, for some descriptors this number may not be practical. Some descriptors may be subject to genotype by environment interaction, and it may be necessary to evaluate the traits in multiple environments. It is important that descriptor data be collected in the format approved by the germplasm system to insure that descriptor data is useful to the user community.

The data for some germplasm descriptors may be quantitative whereas other descriptors may be qualitative in nature. When accessions are evaluated, new single gene markers may also be identified. A listing of previously identified genes in red clover (Quesenberry et al., 1991), and a discussion of procedures for listing newly identified genes is presented in Chapter 3, Reproductive Biology, Genetics and Evolution.

It is also important that unique breeding lines and ploidy levels of red clover and related species be preserved in gene banks. Examples of these types of material include the "stoloniferous" population of red clover recently described from Australia (Smith and Bishop, 1993). The cultivar Astred was selected from CPI 48825 which was originally collected near Crato, Portugal in 1968. Plants of this cultivar are reported to produce strong stolons in late summer with daughter plants appearing within 100mm of the parent. These daughter plants then develop a root system and the connection to the parent plant degenerates. This trait results in superior persistence under grazing compared to other red clover cultivars. Estimated percentage ground cover of Astred three years after seeding was 53 compared to 15 or less for other adapted cultivars. As discussed in Chapter 2, ploidy level manipulations may be important to successful production of interspecific hybrids with red clover. Although numerous tetraploid lines of red clover are currently available in gene banks, a need exists for preservation of tetraploids of related species. Recently, five tetraploid genetic stock of *Trifolium* species have been released and deposited in the U.S. NPGS (Mundell and Taylor, 1995). These include tetraploids of *T. diffusum* Ehrh., *T. pallidum* Bory & Kit. *T. cherleri* L., *T. incarnatum* L., and *T. pauciflorum* Lojac. These lines were produced by use of either nitrous oxide or colchicine.

References

Baker, E., M. Collins and N.L. Taylor. 1992. Developmental patterns and hay drying effects of red clover stem trichomes. *Amer. For. Grassl. Council Proc. Ames, IA*. pp. 235–239.

Banciu, T. and I. Tucra. 1982. Local populations, families and varieties of leguminous and graminaceous perennial forage crops studied at the Turda Agricultural Research Station. (In Romanian) *Contributii Ale Cercetarii Stiintifice la Dezvoltarea Agriculturii Bucharest, Romania*. pp. 481–500.

Bettencourt, E., T. Hazekamp and M.C. Perry. 1992. Directory of Germplasm Collections 7. Forage Legumes, Grasses, Browse Plants, and Others. IBPGR. Rome.

Bowley, S. R., N.L. Taylor and P.L. Cornelius. 1984. Phenotypic recurrent selection for stem length in Kenstar red clover. *Crop Sci.* 24:578–582.

Brown, A.H.D. 1989a. Core collections: A practical approach to genetic resources management. *Genome.* 31:818–824.

Brown, A.H.D. 1989b. The case for core collections pp. 136–156. In A.D.H. Brown, O.H. Frankel, D.R. Marshall and J.T. Williams (eds.), *The Use of Plant Genetic Resources.* Cambridge University Press, Cambridge.

Brown, A.H.D. and J.D. Briggs. 1991. Sampling strategies for genetic variation in *ex situ* collections of endangered plant species. pp. 30–44. In D.A. Falk and K.E. Holsinger (eds.), *Genetics and Conservation of Rare Plants.* Oxford University Press. New York.

Call, Neysa M. and K.H. Quesenberry. 1994. Effect of three genotypes of *Trifolium* on the fecundity of *Meloidogyne arenaria* and *Meloidogyne javanica*. p. 215, In *Agronomy abstracts*. ASA-CSSA-SSSA Madison, WI.

Drach, N.P., L.P. Shcherbina, Yu.S. Bykhatskii, S.I. Pilpenko, N.S. Nazarova and V.M. Rogovchenko. 1988. Use of the world collection of the Vavilov Institute of Plant Industry (VIR) in breeding early red clover. (In Russian) *Sbornik Nauchnykh Trudov po Prikladnoi Botanike Genetike i Selektsii*. 120:64–67.

Drozdova, O.L. 1988. Source material of red clover for breeding for resistance to diseases in the north-west region of the RSFSR. (In Russian) *Sbornik Nauchnykh Trudov po Priklandnoi Botanike Genetike i Selektsii*. 120:67–70.

Forde, M.B. and H.S. Easton. 1986. Report of New Zealand-France-IBPGR forage germplasm collecting expedition to southwestern Europe. Grasslands Division, DSIR, Palmerston North, New Zealand.

Frankel, O.H. 1984. Genetic perspectives of germplasm conservation. pp. 161–170. In W. Arber, K. Llimensee, W.J. Peacock and P. Starlinger (eds.), *Genetic Manipulation: Impact on Man and Society*. Cambridge University Press, Cambridge.

Gillett, J.M. and R.R. Smith. 1985. Germplasm exploration and preservation. In N.L. Taylor (ed.), *Clover Science and Technology*. ASA-CSSA-SSSA Mono. 25:445–456 Madison, WI.

Harlan, J.R. 1975. Crops and Man. *American Society of Agronomy, Madison*, WI.

Hawkes, J.G. 1980. *Crop Genetic Resources Field Collection Manual*. IBPGR, Rome.

Huenneke, L.F. 1991. Ecological implications of genetic variation in plant populations. pp. 30-44. In D.A. Falk and K.E. Holsinger (eds.), *Genetics and Conservation of Rare Plants*. Oxford University Press. New York.

IBPGR. 1984. Forage Legume Descriptors. In S. Anderson and W. Ellis Davies (eds.), IBPGR, Rome.

IBPGR. 1991. *A Guide to European Forage Databases*. European Cooperative Programme for Crop Genetic Resources Networks. IBPGR, Rome.

IPGRI. 1993. Annual Report 1993. International Plant Genetic Resources Institute, Rome 100.

Kouame, C.N. and K.H. Quesenberry. 1993. Cluster analysis of a world collection of red clover. *Genet. Res. & Crop Evol.* 40:39–47.

Kresovich, S., R.L. Jarret, J.R. McFerson, A.K. Szewc-Mcfadden, S.M. Bliek, Z.W. Liu, S.M. Brown, T.Y. Wang and C.J. Hernandez. 1994. Characterization of variation of plant genetic resources employing simple sequence repeat DNA analysis. pp. 219. In *Agronomy abstracts*. ASA-CSSA-SSSA Madison, WI.

Kuleshov, G.F., V.A. Katkov and N.I. Vavilov. 1987. Vavilov's concepts in contemporary herbage crop breeding. (In Russian) *Selektsiya i Semenovodstvo Moscow*. 6:52–54.

Mundell, R.E. and N.L. Taylor. 1995. Registration of four tetraploid genetic stocks. *Crop Sci.*: in press.

Pederson, G.A. and K.H. Quesenberry. 1994. Collecting *Trifolium* germplasm in southwestern Bulgaria. pp. 36–42. In P.R. Narasimhala (ed.), Proc. *Thirteenth Trifolium Conf.*, Charolettetown, Prince Edward Island, Canada. 22–24 June 1994.

Piper, C.V. 1924. *Forage Plants and Their Culture*. MacMillan, New York.

Porceddu, E. and A.B. Dowania. 1992. Sampling variation in genetic resources of seed crops: a review. *Genet. Res. Crop Evol.* 39:39–49.

Quesenberry, K.H., G.M. Prine, O.C. Ruelke, L.S. Dunavin and P. Mislevy. 1993. Registration of 'Cherokee' red clover. *Crop Sci.* 33:208–209.

Quesenberry, K.H. and G.R. Smith. 1990. Exploration for *Trifolium* germplasm in Bulgaria in July 1990. *Progress Report Clovers and Special Purpose Legumes Research*. 23:50–54.

Quesenberry, K.H., R.R. Smith, N.L. Taylor, D.D. Baltensperger and W.A. Parrott. 1991. Genetic nomenclature in clovers and special purpose legumes. I. Red and white clover. *Crop Sci.* 31:861–867.

Rincker, C.M. 1974. Effect of frequent thawing on viability of red clover seed in cold storage. *Crop Sci.* 14:749–750.

Rincker. C.M. 1983. Germination of forage crop seeds after 20 years of subfreezing storage. *Crop Sci.* 23:229–231.

Ryabova, I.V. 1986. Useful breeding forms of red clover in Vologda province. (In Russian) *Nauchno Tekhnicheskii Byulleten Vsesoyuznogo Ordena Lenina i Ordena Druzhby Narodov Nauchno Issledovatel'skogo Instituta Rastenievodstva Imeni N.I. Vavilova.* 165:27–29.

Smith, R.S. and D.J. Bishop. 1993. Astred - a stoloniferous red clover. In *Proc. Intern. Grassl. Congr.* 17:421–423. Palmerston North, Hamilton, Lincoln, New Zealand and Rockhampton, Australia.

Smith, G.R. and C.S. Hoveland. 1994. Forage legume collection in the Republic of Georgia. pp. 43–47. In P.R. Narasimhala (ed.), *Proc. Thirteenth Trifolium Conf.,* Charolettetown, Prince Edward Island, Canada. 22–24 June 1994.

Steiner, J.J. and C.J. Poklemba. 1994. *Lotus corniculatus* classification by seed globulin polypeptides and relationship to accession pedigrees and geographic origin. *Crop Sci.* 34:255–264.

Taylor, N.L. 1973. Red clover and alsike clover. In M.R. Heath, D.S. Metcalfe and R.F. Barnes (eds.), *Forages the Science of Grassland Agriculture,* third edition. The Iowa State University Press. Ames, IA.

Taylor, N.L. 1985. Clovers around the world. In N.L. Taylor (ed.), *Clover Science and Technology.* ASA-CSSA- SSSA, Madison, WI.

Taylor, N.L. and M.D. Rumbaugh. 1986. Survey and collection of *Trifolium* germplasm in Romania, 1984. pp. 47–48. In *Proc. 9th Trifolium Conference.* Guelph, Ontario, Canada. 4–5 June 1986.

Taylor, N.L. and R.R. Smith. 1990. Collection and identification of *Trifolium* germplasm in Yugoslavia. p. 42. In *Proc 11th Trifolium Conference.* Sublimity, Oregon. 9–11 July, 1990.

Taylor, N.L., P.L. Cornelius and M.G. Long. 1985. Phenotypic recurrent selection for multiple-parted heads in red clover. *Crop Sci.* 25:489–494.

Taylor, N.L., P.B. Gibson and W.E. Knight. 1977. Genetic vulnerability and germplasm resources of the true clovers. *Crop Sci.* 17:632–634.

Zohary, M. and D. Heller. 1984. *The Genus Trifolium.* The Israel Academy of Sciences and Humanities. 606 pp.

17. Future Trends

As pointed out in Chapter 1, Historical Perspectives, red clover use has declined in the United States from about 14 million hectares (1927) to 4.15 million hectares (1987). Similar declines have been reported for regions of western Europe, New Zealand, and Australia, and probably others. Of the countries surveyed, only Canada has not suffered a similar decline (See Chapter 15, Seed Production).

The general decline in red clover usage was apparent to Fergus and Hollowell (1960) who suggested that low nitrogen fertilizer prices were a primary cause. This hypothesis is borne out by the study of Allison and Ott (1987) on the economics of using legumes as a nitrogen source in conservation tillage systems. They concluded that nitrogen fertilizer prices would have to increase substantially for legume cover crops to become cost-effective N sources. This, of course, ignores the value of forage for livestock feed, and the prevention of soil erosion. The data cited in Chapter 11, Rhizobium Relationships, support the hypothesis that cheap nitrogen prices are a major factor in the decline of red clover hectarage.

However, another factor has emerged in the United States and probably other countries: the clover-small grain row crop rotation system once so prevalent, has all but disappeared and has been replaced by a soybean-corn system, or a monoculture system. Red clover is now sown more frequently in grass sods for pasture renovation than for rotations. This change has contributed to declines in the use of other legumes as well. Sweet clover (*Melilotus* sp), crimson clover (*Trifolium incarnatum* L.), and hairy vetch (*Vicia villosa* Roth.) have been reduced in area, and scientists who once studied these crops have retired or have been reassigned.

Monoculture or soybeans-corn rotation has been encouraged by government policies. Corn, and other feed grains, wheat, cotton and soybeans received three-fourths of all crop subsidies. This policy provided powerful incentives for farmers to adopt continuous monoculture and to increase their per hectare yields (Parr et al., 1990).

1. Consequences of the decline in use of red clover and other legumes

1.1. *Shift in area of seed production*

Seed production of red clover and other legumes as well as many forage grasses in the U.S. has shifted from the area where legumes are utilized to the Pacific Northwest. Part of the rationale for this shift is the higher seed yields obtained in western U.S. Another major factor for red clover at least, is that without livestock on midwest farms, forage is not needed and the opportunity for seed production, usually a by-product of forage production, no longer exists (Parker, 1990). Also, these above factors predicated a reduction in total quantity of seed produced. Legume seed production in the U.S. (not including alfalfa) dropped from 122 million kg (1959) to 22 million kg (1979) (Power and Doran, 1984). At the same time total nitrogen applied in U.S. increased from 2.4 million tons to 95 million tons. In many areas of the midwest, livestock were moved off former crop rotation farms, fences were removed, and fields have been consolidated to accommodate large machinery.

1.2. *Soil nutrient management*

Because no animals are now on many farms, nutrient recycling has been greatly reduced. Nutrients are removed from farms as crop products, and more fertilizers must be purchased to make up the difference (King, 1990). Likewise, dentrification becomes more of a problem due to lack of red clover and other legumes to serve as a nitrogen sink. Leaching, particularly of nitrogen, is prevalent (Elliott et al., 1987; Scott and Burt, 1987).

1.3. *Soil erosion and tilth*

Erosion of soil under monoculture has been adequately documented (Hargrove and Frye, 1987). Recognition of soil losses has led to minimum tillage practices wherein clover and other legumes provide cover, leading to a decrease in soil and nutrient loss from erosion (Wall et al., 1991). Reductions of rainfall runoff from corn intercropped with red clover vs silage corn ranged from 45 to 87% and soil loss reductions ranged from 46 to 78%. Where no additional N was added, yields from the corn/clover system consistently exceeded yields from the corn silage system. Increasing concerns about soil degradation with continuous corn have led to experiments to determine the efficacy of using red clover and other underseeded crops in rotation. Not only did use of legumes (harvested for hay) increase the yield of corn (1.5 to 8.9% depending upon the year) but it also resulted in a finer seed bed and increased wet aggregate stability compared to continuous corn (Raimbault and Vyn, 1991). Similar results were obtained in an experiment on splash detachment of soil. Splash detachment rates were 48, 40, 39, and 31 mg per cm^3 for continuous corn, corn after soybeans, soybeans after corn, and oats-clover treatments, respectively. Soil aggregation as inferred from geometric mean diameter followed the same trend (Hussain et al., 1988).

1.4. *Water use efficiency and quality*

Precipitation use efficiency is critical in dryland areas. A comparison was conducted in Nebraska (USA) on monoculture and rotation system effects on precipitation use efficiency in corn (Varvel, 1994). A 4-year rotation of oat (*Avena sativa* L.) and clover (80% sweet clover and 20% red clover), grain sorghum, soybeans and corn was used. Precipitation use efficiency was greater in rotation (101.8 kg/ha/cm) than in continuous corn (83.6 kg/ha/cm). The authors concluded that in dryland production areas, cropping systems with greater and more stable precipitation use efficiency (as with clover rotation) can reduce crop failures.

Water quality may also be affected. A more dispersed pattern of livestock production is a necessary step to reduce the often severe regional water quality problems that result when available local supply of manure exceeds the capacity of cropland to use the nutrients in manure efficiently (Benrook, 1990). The potential of N leaching in the groundwater may be greater when N is applied in fertilizer form as compared to N applied by legume cover crops (Hargrove, 1991).

1.5. *Pest problems*

Weed control has become more difficult leading to greater use of herbicides, through development of herbicide resistance in weeds, and greater opportunity for buildup of specific weeds in monoculture systems. An example is the increase in Johnson grass (*Sorghum halapense* L.) and the concurrent greater use of herbicides to combat it (Francis and Clegg, 1990). Insect and disease control also becomes more difficult in monoculture systems (Luna and House, 1990; Byers and Stromberg 1987).

2. Conclusions

The problems of omitting legumes including red clover from a rotation are being recognized. Several symposia have been held to address the problem (Edwards et al., 1990). These symposia include the following terms in their titles: sustainable agriculture, conservation tillage, minimum tillage, regenerative agriculture, and organic farming.

Thus the farming and related communities are becoming aware of the problem. However, government programs will have to be modified so that credit is given for the planting of legumes. Fortunately, some evidence of a shift in this direction is apparent in the government programs in the United States and other countries. Much more remains to be accomplished.

In addition to increased interest in the use of forage legumes to satisfy environmental concerns, red clover as a species is vastly changed from the earlier common types. More persistent cultivars have been developed that possess greater adaptation to the area of use. All major diseases, and nematodes are under control by cultivars bred for resistance. It also may be possible to breed for increased plow down value (as a nitrogen source) (Christie et al., 1992). As a consequence, the use of red clover has been expanded to include dryer as well as more humid areas. Although use of red clovers in rotations with other crops has declined, use in pasture renovation and direct seedings for forage has increased. Cultivars have been developed for use under grazing conditions (i.e. Astred in Tasmania). Also, long range interspecific hybridization programs hold promise for incorporating even greater perenniality into red clover from zigzag clover. Further, the movement of seed production from the area of forage use to specialized areas of seed production has positive aspects. As a consequence of the change from common types produced as a by-product of hay production to improved cultivars, seed production has shifted to western specialized areas in the U.S. This suggests that more seed of improved cultivars and less seed of common types may be available than in the past. Accordingly, in the U.S. at least, overall production has not declined as much as area of usage (Taylor and Smith, 1995).

New legumes are also becoming available, i.e. kura or caucasian clover (*T. ambiguum* Bieb.) and rhizomatous birdsfoot trefoil (*Lotus corniculatus* L.) that undoubtedly will attain a position in forage and animal agriculture. However, red clover, because of its unique qualities of high seedling vigor, ease of establishment and competitiveness will continue to be a major legume in most forage producing countries of the world.

References

Allison, J.R. and S.L. Ott. 1987. Economics of using legumes as a nitrogen source in conservation tillage systems. In J.F. Power (ed.). *The Role of Legumes in Conservation Tillage Systems.* pp. 145–150. *Conf. Proc., Athens, GA. Soil Cons. Soc. Amer.*

Benbrook, C.M. 1990. Society's stake in sustainable agriculture. In Edwards, C.A., R. Lal, P. Madden, R.H. Miller and G. House (eds.), *Sustainable Agricultural Systems.* pp. 68–76. Soil and Water Cons. Soc., Ankeny, IA.

Byers, R.A. and E.L. Stromberg. 1987. Influence of legumes in insects and diseases in conservation tillage systems. In J.F. Power (ed.), *The Role of Legumes in Conservation Tillage Systems.* pp. 1–5. Conf. Proc., Athens, GA. Soil Cons. Soc. Amer.

Christie, B.R., E.A. Clark and R.S. Fulheison. 1992. Comparative plow down value of red clover strains. *Canad. J. Plant Sci.* 72:1207–1213.

Edwards, C.A., R. Lal, P. Madden, R.H. Miller and G. House (eds.). 1990. *Sustainable Agriculture Systems.* 696 pp. Soil and Water Cons. Soc., Ankeny, IA.

Elliott, L.F., R.I. Papendich and D.F. Bezdiceh. 1987. Cropping practices using legumes with conservation tillage and soil benefits. In J.F. Power (ed.) *The Role of Legumes in Conservation Tillage Systems.* pp. 81–89. Conf. Proc., Athens, GA. Soil Cons. Soc. Amer.

Fergus, E.N. and E.A. Hollowell. 1960. Red clover. *Advances in Agron.* 12:365–436.

Francis, C.A. and M.D. Clegg. 1990. Crop rotations in sustainable production systems. In Edwards, C.A., R. Lal, P. Madden, R.H. Miller and G. House (eds.), *Sustainable Agricultural Systems.* pp. 107–122. Soil and Water Cons. Soc., Ankeny, IA.

Hargrove, W.L. and W.W. Frye. 1987. The need for legume cover crops in conservation tillage production. In. J.F. Power (ed.), *The Role of Legumes in Conservation Tillage Systems.* pp. 1–5. Conf. Proc., Athens, GA. Soil Cons. Soc. Amer.

Hargrove, W.L. (ed.), 1991. *Cover Crops for Clean Water.* Soil and Water Conserv. Soc. Ankeny, IA. 198 pp.

Hussain, S.K., L.N. Millke and J. Skopp. 1988. Detatchment of soil as affected by fertility management and crop rotations. *Soil Soc. Amer. J.* 152:1463–1468.

King, L.D. 1990. Soil nutrient management in the United States. In Edwards C.A., R. Lal, P. Madden, R.H. Miller and G. House (eds.), *Sustainable Agricultural Systems.* pp. 89–106. Soil and Water Cons. Soc., Ankeny, IA.

Luna, J.M. and G.J. House. 1990. Pest management in sustainable agricultural surplus. In Edwards, C.A., R. Lal, P. Madden, R.H. Miller and G. House (eds.), *Sustainable Agricultural Systems.* pp. 157 173. Soil and Water Cons. Soc. Ankeny, IA.

Parker, C.F. 1990. Role of animals in sustainable agriculture. In Edwards, C.A., R. Lal, P. Madden, R.H. Miller and G. House (eds.), *Sustainable Agricultural Systems.* pp. 238–245. Soil and Water Cons. Soc. Ankeny, IA.

Parr, J.F., R.I. Papendeck, I.G. Youngberg and R.E. Meyer. 1990. In Edwards, C.A., R. Lal, P. Madden, R.H. Miller and G. House (eds.), *Sustainable Agricultural Systems.* pp. 50–67. Soil and Water Cons. Soc. Ankeny, IA.

Power, J.F. and J.W. Doran. 1984. Nitrogen use in organic farming. In R.D. Hauch (ed.), pp. 585–593. *Nitrogen in Crop Production.* Proc. Symp. May 25–27, 1982. Sheffield, AL, TVA. ASA, Madison, WI.

Raimbault, B.A. and T.J. Vyn. 1991. Crop rotation and tillage effects on corn growth and soil structural stability. *Agron. J.* 83:979–985.

Scott, T.M. and R.F. Burt. 1987. Use of red clover in corn polyculture systems. In J.F. Power (ed.), *The Role of Legumes in Conservation Tillage Systems.* pp. 101–103. Conf. Proc., Athens, GA. Soil Cons. Soc. Amer.

Taylor, N.L. and R.R. Smith. 1995. Red clover. In Barnes, R.F., D.A. Miller and C.J. Nelson. (eds.), Forages, an introduction to grassland agriculture. Fifth Edit. pp. 217–226. Iowa State Univ. Press, Ames, IA.

Varvel, G.E. 1994. Monoculture and rotation system effects on precipitation use efficiency of corn. *Agron. J.* 86:204–208.

Wall, G.J., E.A. Pringle and R.W. Sheard. 1991. Intercropping red clover with corn for silage control. *Canad. J. Soil Sci.* 71:137–145.

About the authors

Norman L. Taylor is professor of Agronomy at the University of Kentucky, where he has served since 1953. He is the author of over 100 refereed journals articles, 16 book chapters, and was editor of the monograph "Clover Science and Technology" published in 1985. He has taught courses on plant breeding and genetics, and forage crops, and conducted research on red clover and related *Trifolium* species during his tenure at the University of Kentucky. He released the Kenstar cultivar and 25 clover germplasms. Dr. Taylor participated in formal USDA, Plant Introduction Service clover explorations in Romania, Yugoslavia, and western USA and has collected clover seeds during travels to France, Sweden, Norway, New Zealand, Australia, Scotland, England, Japan, and in almost all of the states of the U.S. He also collected clovers during sabbaticals at the University of California, Davis; University of Wisconsin, Madison; University of Florida, Gainesville and DSIR, New Zealand. He presently serves as USDA-appointed curator for the genus *Trifolium*, and maintains the world collection of approximately 240 species of *Trifolium*.

Kenneth H. Quesenberry is professor of Agronomy at the University of Florida where he has served since 1975. He is the author of over 50 refereed journal articles and 2 book chapters, and editor of 3 meeting proceedings. He has taught courses in basic genetics, cytogenetics, plant breeding and forage science and range management. He conducted plant improvement research on tropical grasses and legumes as well as red clover and other *Trifolium* species during his career at the University of Florida. He recently released the cultivar Cherokee red clover and a clover germplasm. Dr. Quesenberry participated in formal USDA, Plant Introduction Service clover germplasm explorations to Bulgaria and western USA. He currently serves as chair of the Clover and Special Purpose Legume Crop Advisory Committee of the U.S. National Plant Germplasm System.

Current Plant Science and Biotechnology in Agriculture

16. J.C. Pech, A. Latché and C. Balagué (eds.): *Cellular and Molecular Aspects of the Plant Hormone Ethylene.* 1993 ISBN 0-7923-2169-3
17. R. Palacios, J. Mora and W.E. Newton (eds.): *New Horizons in Nitrogen Fixation.* Proceedings of the 9th International Congress on Nitrogen Fixation (Cancún, Mexico, 1992). 1993 ISBN 0-7923-2207-X
18. Th. Jacobs and J.E. Parlevliet (eds.): *Durability of Disease Resistance.* 1993 ISBN 0-7923-2314-9
19. F.J. Muehlbauer and W.J. Kaiser (eds.): *Expanding the Production and Use of Cool Season Food Legumes.* A Global Perspective of Peristent Constraints and of Opportunities and Strategies for Further Increasing the Productivity and Use of Pea, Lentil, Faba Bean, Chickpea, and Grasspea in Different Farming Systems. Proceedings of the Second International Food Legume Research Conference (Cairo, Egypt, 1992). 1994 ISBN 0-7923-2535-4
20. T.A. Thorpe (ed.): In Vitro *Embryogenesis in Plants.* 1995 ISBN 0-7923-3149-4
21. M.J. Daniels, J.A. Downie and A.E. Osbourn (eds.): *Advances in Molecular Genetics of Plant-Microbe Interactions.* Volume 3. 1994 ISBN 0-7923-3207-5
22. M. Terzi, R. Cella and A. Falavigna (eds.): *Current Issues in Plant Molecular and Cellular Biology.* Proceedings of the VIIIth International Congress on Plant Tissue and Cell Culture (Florence, Italy, 1994). 1995 ISBN 0-7923-3322-5
23. S.M. Jain, S.K. Sopory and R.E. Veilleux (eds.): In Vitro *Haploid Production in Higher Plants.* Volume 1: Fundamental Aspects and Methods. 1996 ISBN 0-7923-3577-5
24. S.M. Jain, S.K. Sopory and R.E. Veilleux (eds.): In Vitro *Haploid Production in Higher Plants.* Volume 2: Applications. 1996 ISBN 0-7923-3578-3
25. S.M. Jain, S.K. Sopory and R.E. Veilleux (eds.): In Vitro *Haploid Production in Higher Plants.* Volume 3: Important Selected Plants. 1996 ISBN 0-7923-3579-1
26. S.M. Jain, S.K. Sopory and R.E. Veilleux (eds.): In Vitro *Haploid Production in Higher Plants.* Volume 4: Cereals. 1996 ISBN 0-7923-3978-9
27. I.A. Tikhonovich, N.A. Provorov, V.I. Romanov and W.E. Newton (eds.): *Nitrogen Fixation: Fundamentals and Applications.* 1995 ISBN 0-7923-3707-7
28. N.L. Taylor and K.H. Quesenberry: *Red Clover Science.* 1996 ISBN 0-7923-3887-1
29. S.M. Jain, S.K. Sopory and R.E. Veilleux (eds.): In Vitro *Haploid Production in Higher Plants.* Volume 5: Oil, Ornamental and Miscellaneous Plants. 1996 ISBN 0-7923-3979-7

KLUWER ACADEMIC PUBLISHERS – DORDRECHT / BOSTON / LONDON

Current Plant Science and Biotechnology in Agriculture

1. H.J. Evans, P.J. Bottomley and W.E. Newton (eds.): *Nitrogen Fixation Research Progress.* Proceedings of the 6th International Symposium on Nitrogen Fixation (Corvallis, Oregon, 1985). 1985 ISBN 90-247-3255-7
2. R.H. Zimmerman, R.J. Griesbach, F.A. Hammerschlag and R.H. Lawson (eds.): *Tissue Culture as a Plant Production System for Horticultural Crops.* Proceedings of a Conference (Beltsville, Maryland, 1985). 1986 ISBN 90-247-3378-2
3. D.P.S. Verma and N. Brisson (eds.): *Molecular Genetics of Plant-microbe Interactions.* Proceedings of the 3rd International Symposium on this subject (Montréal, Québec, 1986). 1987 ISBN 90-247-3426-6
4. E.L. Civerolo, A. Collmer, R.E. Davis and A.G. Gillaspie (eds.): *Plant Pathogenic Bacteria.* Proceedings of the 6th International Conference on this subject (College Park, Maryland, 1985). 1987 ISBN 90-247-3476-2
5. R.J. Summerfield (ed.): *World Crops: Cool Season Food Legumes.* A Global Perspective of the Problems and Prospects for Crop Improvement in Pea, Lentil, Faba Bean and Chickpea. Proceedings of the International Food Legume Research Conference (Spokane, Washington, 1986). 1988 ISBN 90-247-3641-2
6. P. Gepts (ed.): *Genetic Resources of* Phaseolus *Beans.* Their Maintenance, Domestication, Evolution, and Utilization. 1988 ISBN 90-247-3685-4
7. K.J. Puite, J.J.M. Dons, H.J. Huizing, A.J. Kool, M. Koorneef and F.A. Krens (eds.): *Progress in Plant Protoplast Research.* Proceedings of the 7th International Protoplast Symposium (Wageningen, The Netherlands, 1987). 1988 ISBN 90-247-3688-9
8. R.S. Sangwan and B.S. Sangwan-Norreel (eds.): *The Impact of Biotechnology in Agriculture.* Proceedings of the International Conference The Meeting Point between Fundamental and Applied in vitro Culture Research (Amiens, France, 1989). 1990. ISBN 0-7923-0741-0
9. H.J.J. Nijkamp, L.H.W. van der Plas and J. van Aartrijk (eds.): *Progress in Plant Cellular and Molecular Biology.* Proceedings of the 8th International Congress on Plant Tissue and Cell Culture (Amsterdam, The Netherlands, 1990). 1990 ISBN 0-7923-0873-5
10. H. Hennecke and D.P.S. Verma (eds.): *Advances in Molecular Genetics of Plant–Microbe Interactions.* Volume 1. 1991 ISBN 0-7923-1082-9
11. J. Harding, F. Singh and J.N.M. Mol (eds.): *Genetics and Breeding of Ornamental Species.* 1991 ISBN 0-7923-1094-2
12. J. Prakash and R.L.M. Pierik (eds.): *Horticulture – New Technologies and Applications.* Proceedings of the International Seminar on New Frontiers in Horticulture (Bangalore, India, 1990). 1991 ISBN 0-7923-1279-1
13. C.M. Karssen, L.C. van Loon and D. Vreugdenhil (eds.): *Progress in Plant Growth Regulation.* Proceedings of the 14th International Conference on Plant Growth Substances (Amsterdam, The Netherlands, 1991). 1992 ISBN 0-7923-1617-7
14. E.W. Nester and D.P.S. Verma (eds.): *Advances in Molecular Genetics of Plant–Microbe Interactions.* Volume 2. 1993 ISBN 0-7923-2045-X
15. C.B. You, Z.L. Chen and Y. Ding (eds.): *Biotechnology in Agriculture.* Proceedings of the First Asia-Pacific Conference on Agricultural Biotechnology (Beijing, China, 1992). 1993 ISBN 0-7923-2168-5